An Introduction to Recombinant DNA Techniques

Basic Experiments in Gene Manipulation

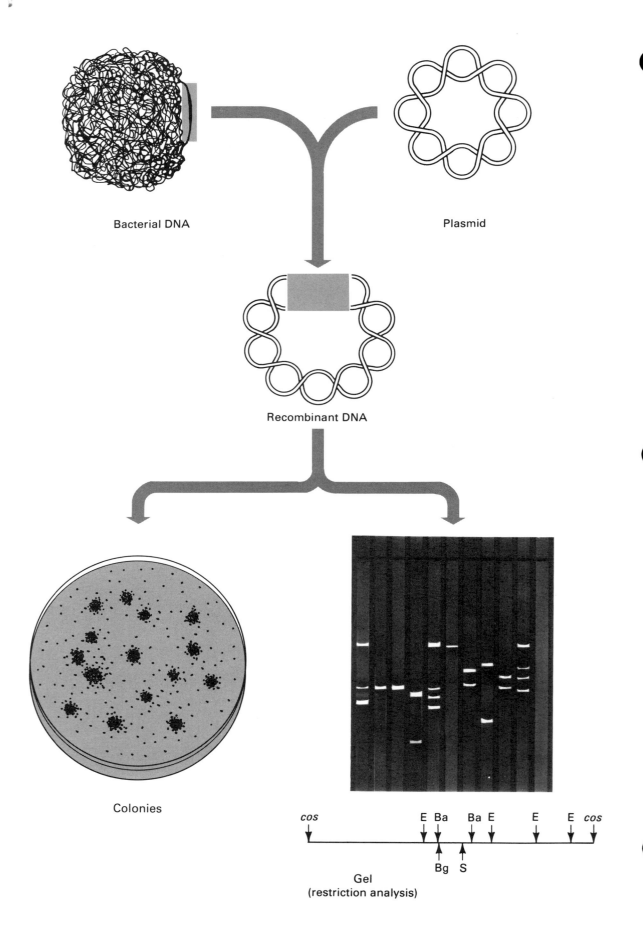

Bacterial DNA

Plasmid

Recombinant DNA

Colonies

Gel
(restriction analysis)

cos   E Ba   Ba E   E   E *cos*

Bg   S

An Introduction to Recombinant DNA Techniques

Basic Experiments in Gene Manipulation

Perry B. Hackett

James A. Fuchs

Joachim W. Messing

University of Minnesota

The Benjamin/Cummings Publishing Company, Inc.
Menlo Park, California • Reading, Massachusetts
London • Amsterdam • Don Mills, Ontario • Sydney

Sponsoring Editor: Jane R. Gillen
Production Coordinator: Greg Hubit
Copy Editor: Debra Myson-Etherington
Book Designer: Susan Rogin
Cover Design: Henry Breuer
Artists: Scientific Illustrators and Kris Kohn

Library of Congress Cataloging in Publication Data

Hackett, Perry B.
    An introduction to recombinant DNA techniques.

    (Benjamin/Cummings series in the life sciences)
    Includes bibliographical references.
    1. Recombinant DNA—Experiments. 2. Recombinant DNA—
Laboratory manuals. 3. Genetic engineering—Experiments.
4. Genetic engineering—Laboratory manuals. I. Fuchs,
James A. II. Messing, Joachim W. III. Title. IV. Series.
QH442.H33 1984   574.87'3283   83-21421
ISBN 0-8053-3672-9

    defghij-AL-89876

The Benjamin/Cummings Publishing Company, Inc.
2727 Sand Hill Road
Menlo Park, California 94025

The Benjamin/Cummings Series in the Life Sciences

F. J. Ayala
Population and Evolutionary Genetics: A Primer (1982)

F. J. Ayala and J. A. Kiger, Jr.
Modern Genetics, second edition (1984)

F. J. Ayala and J. W. Valentine
Evolving: The Theory and Processes of Organic Evolution (1979)

C. L. Case and T. R. Johnson
Laboratory Experiments in Microbiology (1984)

R. E. Dickerson and I. Geis
Hemoglobin (1983)

P. B. Hackett, J. A. Fuchs, and J. W. Messing
An Introduction to Recombinant DNA Techniques: Basic Experiments in Gene Manipulation (1984)

L. E. Hood, I. L. Weissman, W. B. Wood, and J. H. Wilson
Immunology, second edition (1984)

J. B. Jenkins
Human Genetics (1983)

K. D. Johnson, D. L. Rayle, and H. L. Wedberg
Biology: An Introduction (1984)

R. J. Lederer
Ecology and Field Biology (1984)

A. L. Lehninger
Bioenergetics: The Molecular Basis of Biological Energy Transformations, second edition (1971)

S. E. Luria, S. J. Gould, and S. Singer
A View of Life (1981)

E. N. Marieb
Human Anatomy and Physiology Lab Manual: Brief Edition (1983)

E. N. Marieb
Human Anatomy and Physiology Lab Manual: Cat and Fetal Pig Versions (1981)

E. B. Mason
Human Physiology (1983)

A. P. Spence
Basic Human Anatomy (1982)

A. P. Spence and E. B. Mason
Human Anatomy and Physiology, second edition (1983)

G. J. Tortora, B. R. Funke, and C. L. Case
Microbiology: An Introduction (1982)

J. D. Watson
Molecular Biology of the Gene, third edition (1976)

W. B. Wood, J. H. Wilson, R. M. Benbow, and L. E. Hood
Biochemistry: A Problems Approach, second edition (1981)

Preface

In the past decade, the use of genetic engineering has spread from university research laboratories to industrial laboratories and, just recently, back into the curricula of college biology departments. For the past several years, we have taught at the University of Minnesota a laboratory course in recombinant DNA techniques, a ten-week course with two formal laboratory sessions per week. Because the theoretical foundations for the recombinant DNA methodologies are in principle quite simple, the course prerequisites do not go beyond a good college background in biology, chemistry, and genetics. Enrolled in our course have been a diverse group of graduate and advanced undergraduate students, many of whom have come to the course with minimal laboratory experience in molecular biology. We found that the available methods manuals and more descriptive treatises on gene manipulation, while excellent for reference purposes, were not appropriate lab manuals for our course. Accordingly, we developed our own sequence of experiments and a number of simplified protocols. After many revisions and refinements, the result is this book. We hope it will prove useful to other instructors and their students, as well as to scientists who want an efficient way to learn basic recombinant DNA techniques independently.

SCOPE AND SEQUENCE

Part One of this book consists of five chapters that introduce the basic principles of gene cloning, give essential background on working with *E. coli*, and describe the three cloning systems to be used. By reading these chapters, students acquire an understanding of the basic principles of the course without distraction by procedural details.

In Part Two, a unique sequence of carefully designed experiments enables the user of this book to become familiar with a variety of techniques in a relatively short time. In only eighteen experiments (twenty periods), students are taken from a simple entry point—determining the number of viable bacteria in a given volume—through cloning of an *E. coli* gene in the three major types of cloning vectors: λ phage, plasmids, and the single-stranded DNA phage M13. By the time the last experiment (on site-specific mutagenesis) is completed, students have achieved a sophisticated level of laboratory expertise in gene manipulation and a solid understanding of the principles of genetic engineering.

The individual protocols are designed to be effective, reliable, fast, and as inexpensive as possible. Protocols requiring ultracentrifugation have been avoided. By using *E. coli* genes—the genes encoding ribonucleotide reductase, *nrdA* and *nrdB*—the experiments are exempt from federal guidelines concerning recombinant DNA, and there are no problems with biohazards. In addition, the *nrd* genes can be identified by their biological properties, thus obviating the need for radioisotopes. Part Three contains appendices providing detailed information about reagents and their sources, necessary equipment, and some of the key procedures.

MATERIALS FOR THE LABORATORY

Our goal is to make this manual usable in any college teaching laboratory. To help the instructor, we have arranged for Bethesda Research Laboratories (BRL) to assemble reagent packages, which will provide all the specialized materials needed for a class of 20 students. Three separate packages are available: (1) a biological package containing all required bacterial phage strains, enzymes, and nucleic acids; (2) a chemicals package containing antibiotics and ultrapure chemical reagents; and (3) a media package. For further information about the packages, please see Appendix A and contact BRL directly at the address given on the next page.

While we believe there are obvious benefits to using the entire sequence of experiments, we realize that this is not always possible or desirable. Therefore, the BRL biological package will include nucleic acid intermediates that will allow performance of experiments out of sequence.

ACKNOWLEDGMENTS

We acknowledge with gratitude the assistance of numerous people in the development of this book. We are grateful to the members of our laboratories who helped improve the techniques used in this book; in particular, we want to thank Betsy Kren for constructing a number of *E. coli* strains and testing some techniques. We owe a major debt to the students and teaching assistants who participated in the previous offerings of our course at the University of Minnesota; as a result of their suggestions, the final version of this manual is significantly better than the original version. Extremely valuable suggestions were also made by the reviewers of the manuscript: David Freifelder of the University of California, San Diego; Joyce Maxwell of California State University at Northridge; John Reeve of Ohio State University; LeLeng To of Gaucher College; and William Wood of the

University of Colorado. We are especially grateful to David Freifelder, who went over an early draft of the manuscript with a fine-tooth comb.

The artwork in this manual was designed and executed by Kris Kohn and Scientific Illustrators, Inc. We deeply appreciate the help of Janet McCallister and Leslie Maggi, who showed extraordinary patience and understanding during their typing of the many drafts. We are also indebted to Jane Gillen of the Benjamin/Cummings Publishing Company for her efforts and encouragement in directing this project through the long process leading to publication.

Perry B. Hackett
James A. Fuchs
Joachim W. Messing

Reagent packages to accompany this book are available from

Bethesda Research Laboratories
P.O. Box 6009
Gaithersburg, MD 20877

Toll-free phone number: 800-638-4045 (U.S.)

Contents

PART THREE   APPENDICES

Part One
Basic Principles

Chapter 1

Introduction

Recombinant DNA technology has revolutionized molecular biology and genetics. Today, virtually any segment of DNA, the genetic material of all cells and of most viruses, can be isolated and replicated to provide sufficient quantities of genes to study their structure and expression. Furthermore, cellular systems can be designed to produce large quantities of particular biological substances. Recently, recombinant DNA techniques have been used in new industrial and medical ventures to produce economically important substances of high purity.

The purpose of this laboratory course is to introduce you to several of the many techniques for gene cloning using three different types of carrier DNA molecules. These carrier molecules, which are called *cloning vectors*, are used to introduce DNA fragments into cells for replication and amplification. The process of insertion followed by establishment of the hybrid vector in a cell is called *molecular cloning* or, simply, *cloning*. In this manual, we will focus on the three cloning vectors most frequently used—double-stranded DNA genomes of viruses, single-stranded DNA genomes of viruses and plasmids. An understanding of the properties of these vectors is essential for anyone interested in using molecular cloning to obtain various biological products. In the exercises described in this course, two genes of the bacterium *Escherichia coli* (*E. coli*)—*nrdA* and *nrdB*, which encode the B1 and B2 subunits of ribonucleotide reductase and which have already been cloned in bacteriophage λ—will be sequentially transferred (subcloned) from the double-stranded DNA of phage λ to the plasmid pBR325, and then from the plasmid DNA into the single-stranded DNA of phage M13.

We will present methods for growing and isolating these cloning vectors, for cleaving DNA into fragments that can be moved from one cloning vector to the next, and for selecting specific recombinant DNA molecules.

3

As you move the genes from one cloning vector to the next, you will analyze recombinant DNA molecules by several techniques—biological tests for gene complementation and drug resistance, physical tests for size determinations by electrophoresis, and mapping of restriction-endonuclease cleavage sites.

A summary of the course is presented as a flowchart in Figure 1-1. This complex figure, with symbols for various recombinant DNA molecules and vectors, should be used throughout the course as a map to locate where you are at any particular time and to see where you are going. You will appreciate the figure more fully when you have completed Part One of the manual.

You may wish to consult textbooks of molecular biology, microbiology, and biochemistry for help in understanding the fundamental concepts underlying the experiments and procedures used in this manual. In addition, we recommend that you obtain a text that covers the principles of cloning. Several useful books and review articles are listed in the references at the end of this manual. Furthermore, you can obtain—free of charge—informative catalogues from companies selling reagents and materials used in gene cloning. Some companies are Bethesda Research Laboratories, Boehringer Mannheim Biochemicals, and New England Biolabs (see Appendix A). These catalogues contain a wealth of practical information not presented in this manual.

This book is divided into three major sections. The first section consists of five chapters that contain a brief discussion of the microbiological techniques required in the course and descriptions of the three cloning systems to be used. Double-stranded DNA vectors—in particular, plasmids and phage λ—have been used for gene cloning for the past several years, and an extensive literature on their use is available. The employment of single-stranded phage DNA as a cloning vector is quite recent so there are fewer publications to which you can refer. Thus, the discussion of M13 will be more extensive than that of plasmid and phage λ DNA.

Figure 1-1 Flowchart of experiments described in this manual. The chart shows the progressive subcloning of the *E. coli* *nrdA* and *nrdB* genes, which are abbreviated *nrd*. Initially, a recombinant λ phage lysogen that contains the *nrd* genes, designated λd*nrd*⁺, is induced, the phage is harvested, and the *nrd* genes are recombined with plasmid pBR325 to form the recombinant plasmid pBR*nrd*. The *nrd* genes are then further subcloned into the single-stranded phage M13 to form the recombinant phage M13*nrd*. In addition, an experiment illustrating the technique of site-specific mutagenesis is shown in the lower right corner of the figure. DNA molecules cleaved by specific restriction endonucleases are designated by (DNA) X (endonuclease); thus, λ phage cleaved by the restriction enzyme *Eco*RI is designated λ X *Eco*RI. The numbers in parentheses indicate the laboratory period during which the particular step is accomplished. The letters in parentheses indicate the procedures used: (a) biological selection or analysis, (b) induction of phage, (c) isolation of DNA, (d) restriction endonuclease cleavage of DNA, (e) ligation of DNA fragments, (f) transformation of *E. coli* cells with recombinant DNAs, (g) electrophoretic analysis of DNAs.

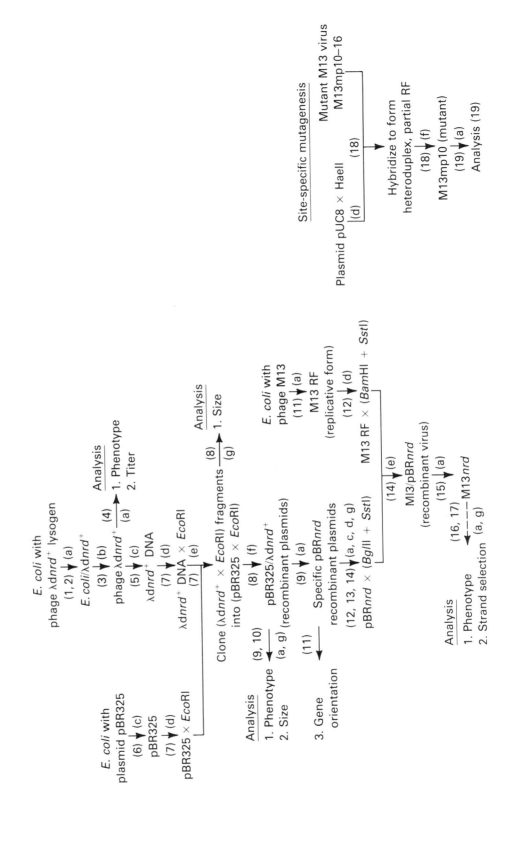

Part Two of this manual is the core of the course. It contains the detailed procedures required to transfer genes from one vector to another. Protocols are presented for 20 laboratory sessions. In previous courses, we have found that students working in pairs need about four hours per session, if they come to the laboratory prepared.

Part Three contains appendices that provide detailed coverage of many of the subjects considered in the main body of the manual. [The best single source, available at present, for more advanced protocols is *Molecular Cloning* by Maniatis, Fritsch, and Sambrook (1982).] In particular, Appendix A lists the supplies, equipment, and reagents required for the entire course. All chemicals, solutions, growth media, and biological reagents used are numerically designated in the Materials per Team section of each experiment, and are elaborated in Appendix A using the same numerical code.

For some years many concerns have been voiced about the hazards of recombinant DNA research. No hazardous microbiological materials are used in this manual. At irregular intervals, the National Institutes of Health (NIH) issues *Guidelines for Research Involving Recombinant DNA Molecules* in which recombinant DNA is defined as ". . . either (i) molecules which are constructed outside living cells by joining natural or synthetic DNA segments to DNA molecules that can replicate in a living cell, or (ii) DNA molecules that result from the replication of those described in (i) above."

In this course, genes from *E. coli* will be cloned by reintroducing them into other *E. coli* strains; consequently, all experiments in this manual are exempt from restricted use by Provision I-E-3 of the guidelines, which states: "The following recombinant DNA molecules are exempt from these guidelines, and no registration with NIH is necessary . . . I-E-3. Those that consist entirely of DNA from a prokaryotic host, including its indigenous plasmids or viruses, when propagated only in that host (or a closely related strain of the same species) or when transferred to another host by well-established physiological means."

Although the experiments in this manual are exempt from the guidelines, all material used in the laboratory should be handled in a manner that conforms to good microbiological practice. You will be taught these practices during the first periods of the course. In general, the fundamental rule of the laboratory is to use common sense at all times.

You must obtain a laboratory coat and wear it in the laboratory. In addition, you must be aware that some instruments are hazardous (for example, the ultraviolet light source), and that many of the chemicals and reagents may be harmful if ingested (for example, ethidium bromide), or dropped onto exposed skin (for example, phenol). Therefore, you must take care and wear appropriate apparel when necessary—for example, plastic goggles when using ultraviolet light, and gloves when using ethidium bromide and phenol. As far as is known, the biological materials—phages, bacteria, and enzymes—are harmless; nonetheless, we will use weakened strains of *E. coli* to prevent any possible spread of recombinant DNA molecules out of the laboratory.

Chapter 2

Growth and Maintenance of *Escherichia coli*

GENERAL INFORMATION ABOUT *E. COLI*

Most research in molecular genetics has utilized the bacterium *Escherichia coli*. The knowledge accumulated, as well as the techniques developed to manipulate and study genes in *E. coli*, has made this bacterium the natural choice for an organism for recombinant DNA technology. Thus, some understanding of the biology and genetics of *E. coli* is useful before beginning laboratory work. Most of the *E. coli* strains you will encounter and all of those used in this course were derived ultimately from a single parent strain called *E. coli* K12. During the past 30 years, thousands of strains have been made from *E. coli* K12 in different laboratories. These strains are usually designated by one or more letters that indicate the laboratory where the strain was derived and by numbers that usually indicate the temporal order in which the strains were isolated. The pedigrees of many widely used strains of *E. coli* K12 are given in a paper by Bachmann (1972).

All laboratory strains of *E. coli* can grow in chemically defined media consisting of a few inorganic compounds, trace amounts of several ions, an organic compound that can serve as both a source of energy and carbon atoms, and one or more organic compounds that may be required by particular strains. Such a growth medium is called a *minimal medium*. Many different minimal media are used for growing *E. coli*, but the differences are not significant for most purposes. In this course, you will use M9 medium; this medium is notable for its very high concentration of phosphate, which serves both as a source of phosphate and as a buffer to maintain the pH at a desirable value.

Many common laboratory strains possess mutations that cause the strain to require an amino acid, purine, pyrimidine, vitamin, or other nutrient, and to grow these strains a minimal medium must be supplemented with the requisite substances. A strain with such a mutation is called an **auxotroph**. A **prototroph** does not require any organic compound other than a carbon source. For example, a histidine auxotroph requires that histidine be present in the medium, whereas a histidine prototroph does not. A mutation that makes a strain auxotrophic is easily detected by observing the absence of growth when a particular substance is lacking. Such mutations are termed **genetic markers**. To verify that a particular strain or a culture has not become contaminated, genetic markers should be checked.

E. coli can utilize many organic compounds as a source of carbon and energy but glucose is the best source and the one most frequently used in the laboratory. Most strains in common use possess mutations that block the utilization of various sugars or other carbon sources. These mutations can be identified by putting the strains in a minimal medium containing the particular substance as the sole source of carbon atoms and noting whether there is growth. This type of mutation (also called a genetic marker) can be used either for testing the purity of strains or as a selective marker in constructing new strains.

The growth of *E. coli* is accelerated by supplying a culture with amino acids, pyrimidines, purines, and vitamins, because then the cells do not have to expend energy synthesizing these substances. An inexpensive way to supply these compounds is to use an enriched but chemically undefined medium, called either a **rich medium** or a **broth**, prepared by hydrolysis of biological materials, such as milk, yeast, or meat. (Actually, a ready-made powder can be bought.) Yeast extract provides most of these substances; tryptone and peptone are enzymatic digests of milk and meat, respectively, and are very rich in amino acids but lack the other substances. It is interesting that all rich media are not equally effective in producing cells for particular purposes and, for this reason, several different rich media will be used in this course.

The standard nomenclature for genes and genetic markers is that of Demerec et al. (1966). A three-letter italicized or underlined symbol is used to describe a genetic locus—for example, *his* (histidine) or *gal* (galactose). All histidine-requiring mutants are written *his* or *his*⁻. Often there are several enzymes in a locus—for example, several enzymes are needed to synthesize histidine; a deficiency of any of these will be *his*⁻. A capital letter following *his* is used to distinguish the loci—for example, *hisA, hisB,* and so on. Usually, each locus represents a particular polypeptide chain or regulatory element (operator or promoter) in the system. Each independently isolated mutation within a particular locus is given a number, for example, *hisA*38. Not all *hisA* mutations are identical or yield exactly the same phenotype but all *hisA*38 mutations are identical. The phenotype of a cell is written with a capital letter and is not italicized; a haploid cell carrying the *hisA*38 marker has the His⁻ phenotype.

The genotype of a strain is usually given by listing all loci known to be different from "wild type." When necessary, a (+) is appended to a genetic symbol to indicate a wild-type allele of a locus—for example, *hisC*$^+$ means a wild-type *hisC* gene. Sometimes a phage genome will be integrated into a bacterial genome. This state of the phage genome is called a ***prophage*** and the bacterium is called a ***lysogen***. In this state, expression of the phage genome is largely repressed. When a strain is lysogenic for a particular prophage, the prophage genotype is written in parentheses—for example, *nrdAhisC*(λcI857) denotes a bacterium carrying the *nrdA* and *hisC* mutations plus a λ prophage in which there is a mutation named *cI857*. Some phage variants contain *E. coli* genes replacing phage genes; such a variant is called a ***transducing particle***. If the genes replaced are essential, the particle will be unable to form a plaque; the phage is ***defective***, and this is denoted by the letter d preceding the symbol for the bacterial gene. For example, if the *lac* genes replace essential phage genes of λ, the transducing particle would be designated λd*lac*$^+$. If the phage genes that are replaced are not essential, the phage is a ***plaque-former*** and this is denoted by the letter p—a *lac*-containing λ would then be denoted by λp*lac*. DNA from a transducing particle can be a prophage so one might have a bacterium whose genotype is *his*$^-$*lac*$^-$(λd*lac*$^+$); its phenotype would be His$^-$Lac$^+$(λ).

A detailed linkage map of *E. coli* has been published by Bachmann and Low (1980). In addition to the map, this paper is a valuable source of references to original research papers describing mutations found in many strains; it also contains useful references to techniques used in modern microbial genetics. Most strains of *E. coli* that you might want can be obtained free of charge from Dr. Barbara Bachmann, Department of Human Genetics, Yale University School of Medicine, 333 Cedar Street, New Haven, CT 06510.

Three books published by the Cold Spring Harbor Laboratory are valuable sources for additional genetic and recombinant DNA techniques: *Advanced Bacterial Genetics* by Davis, Botstein, and Roth (1980); *Molecular Cloning* by Maniatis, Fritsch, and Sambrook (1982); *Experiments in Molecular Genetics* by Miller (1972).

PURIFICATION OF A BACTERIAL STRAIN

All work in microbiology must be done with pure strains—that is, cultures in which all cells are identical. Since mutations occur and strains occasionally become inadvertently contaminated, a culture taken from storage or received from another laboratory should be purified. This is accomplished by obtaining a single colony from the culture, testing that colony for the desired characteristics, and using that colony to prepare a new culture. To obtain a colony, cells are spread on an agar surface sufficiently so that many single cells are well separated. Each single cell forms a colony containing from 10^6–10^9 cells. A quick way to spread cells is called ***streaking***.

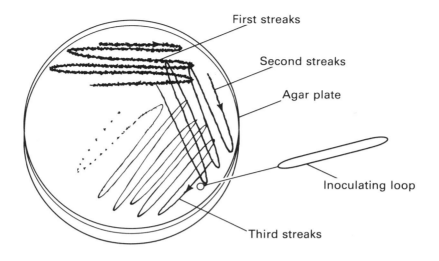

Figure 2-1 Isolation of single colonies by streaking.

First streaks

Second streaks

Agar plate

Inoculating loop

Third streaks

A wire loop is sterilized by placing it in a Bunsen flame until the loop is red-hot. After cooling the wire by touching it to a sterile agar surface for a few seconds, the loop is touched to a solid culture of bacteria or dipped into a liquid culture and then streaked back and forth on agar near the edge of a petri dish (Figure 2-1). The loop is flamed again, cooled as before, and then streaked back and forth again, starting from the edge of the previous streak. This flaming and restreaking is done three times, by which time 50–100 cells should be well separated, and single colonies should result after overnight incubation of the plate. Usually, a rich agar is used, so all cells will grow quickly.

Several single colonies obtained as just described are then tested for genetic characteristics. Testing is done with sets of test plates that are given a reference orientation mark on the bottom and placed on a ruled paper grid such as those shown in Figure 2-2. Two or more test plates are always used. One of these—the master plate—is a rich plate; the other may be a rich plate containing an antibiotic (if antibiotic resistance is to be tested) or a minimal plate that lacks some essential nutrient (if an auxotrophic characteristic is being tested). Using a sterile toothpick, a colony is picked from the streaked plate; then the surface of each plate is touched with the toothpick at corresponding numbered positions. The plates are then incubated until growth occurs. The positions at which growth occurs are noted to determine the genotype of each colony. Usually, ten colonies will be sufficient for this test.

Figure 2-2 Grid patterns.

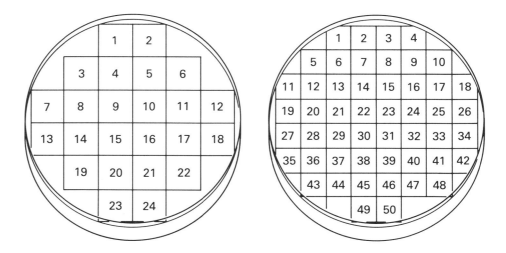

STORAGE OF BACTERIAL CULTURES

Numerous methods are used to store *E. coli* strains. The most permanent, but most time-consuming, method is freeze-drying (*lyophilization*), by which strains have been kept alive for more than 30 years. [See Lapage et al. (1970) for a detailed description of this method.]

A less permanent but very useful method is ultracold storage at $-70°$ C or in liquid nitrogen. To prevent bursting of the cells during freezing, the culture medium must contain either 7% dimethylsulfoxide (DMSO) or 15% glycerol. Cells are removed with a sterile loop from frozen cultures as ice crystals, and either allowed to thaw on an agar plate (and then streaked) or placed in sterile growth medium. If the glycerol content is increased to 40–50%, cultures can be stored in a liquid state for several years at $-20°$ C.

For convenience, stable strains of *E. coli* can be stored at room temperature as a deep stab in airtight tubes containing rich agar. Such stab cultures last for several years. They are prepared by autoclaving a screwcap tube containing liquid agar and allowing the agar to harden. Bacteria are picked with a wire loop from an agar plate or a liquid culture and the loop is stabbed into the rich agar. The tube is then placed in an incubator and the bacteria are grown for 24 hours, after which the tube is sealed with paraffin and stored at room temperature. The lifetime of a deep stab is determined primarily by the quality of the airtight seal, because it is essential that the agar does not dry.

For routine use in the laboratory, a slant culture is used. Liquid agar is placed in a small screw-cap tube and sterilized. The tube is tipped to increase the surface area of the liquid and the agar is allowed to harden. The agar surface is then covered with bacteria by using either a wire loopful of cells or by placing a droplet of culture directly on the surface. The tube is incubated overnight to allow for cell growth and the cap is then tightened. Slant cultures are stored in a refrigerator and will last several months—less if they are opened frequently.

GROWTH OF *E. COLI* IN LIQUID MEDIUM

Growth of a culture usually begins by transferring nongrowing cells—an *inoculum*—from a slant culture (or other storage medium) into a small volume of liquid growth medium. Nongrowing cells do not immediately begin to multiply, because they have adapted to the nongrowth conditions of the slant and need time to synthesize enzymes, ribosomes, and other components required for protein, RNA, and DNA synthesis before growth can begin. This period of nongrowth is called the *lag phase* (Figure 2-3). As growth begins, the culture gradually achieves a state in which all cells are multiplying and in which the cell concentration doubles at a constant rate. In this growth phase, the concentration of the culture increases exponentially, so the phase is called ti. *_xponential phase*. At 37°C, the lag phase is usually 10–60 minutes. The lag phase depends on four factors—the composition of the growth medium, the time the cells have been in stationary phase, the genotype of the bacteria, and the degree of aeration. Exponential growth continues until the composition of the medium changes or oxygen becomes limiting. At this point, growth of the culture slows; when growth ceases, the culture is said to be in the *stationary phase*. After a long period of time in the stationary phase, most cultures enter the *death phase* in which the number of viable cells decreases. Cells in exponential growth may be diluted into fresh medium having the identical composi-

Figure 2-3 Pattern of growth for a population of *E. coli* grown in liquid culture.

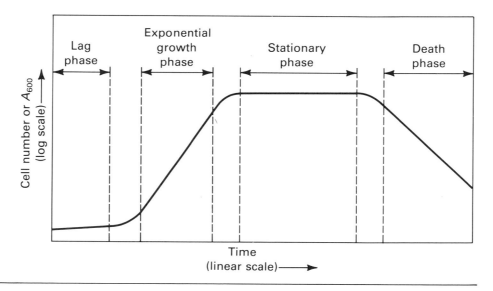

tion without a change in the growth rate and, by continual dilution, an *E. coli* culture can be kept in the exponential phase indefinitely. Also, if an exponentially growing culture is rapidly chilled to below 8°C (by shaking a growth flask in ice water), stored at 4°C, and then rewarmed rapidly to the original temperature at which the culture had been growing, growth will resume without a lag.

At 37°C, a typical *E. coli* strain doubles in cell number in 20 minutes in rich medium and in 50–60 minutes in a minimal medium with glucose as the carbon source; if a slowly utilized carbon source is used, the doubling time will increase severalfold, the factor depending on the carbon source. In all media at 30°C the doubling time is roughly 1.5 times greater than the time at 37°C.

E. coli can grow both aerobically or anaerobically, although anaerobic growth does not occur with all carbon sources. With glucose, anaerobic growth occurs but at about 10% the rate of aerobic growth. At a density of approximately 10^7 cells/ml in liquid medium oxygen cannot diffuse from the atmosphere to the cells fast enough for aerobic growth, and unless the medium is shaken or directly aerated by bubbling, growth of a culture slows considerably at this cell density. Even with aeration, the oxygen supply is inadequate above a cell concentration of 2–3 $\times$ 10^9 cells/ml and few cultures can exceed this cell density.

Since many cells in an inoculum obtained from a slant culture are dead, most experiments begin by inoculating a small volume of growth medium and then growing the cells overnight. In the morning, cell growth will have stopped and the cell density will usually be 2–3 $\times$ 10^9 cells/ml. To obtain actively growing cells for any particular experiment, the overnight culture is diluted 10–100 times and regrown for several hours, at which time the cells are usually in exponential growth.

DETERMINATION OF CELL MASS AND CELL NUMBER

Growth of an *E. coli* culture can easily be monitored by *spectrophotometry*. A spectrophotometer is an instrument that measures the amount of light, at a particular wavelength, that passes through a solution. A beam of light is directed through a bacterial culture, and photons are scattered (deflected) from their original path. The amount of light received by a detector is usually expressed as a logarithmic function of the fraction of the incident light that is *not* received by the detector; this function is called the *absorbance, A*. The wavelength, in nanometers, used for the measurement is written as a subscript; in this course, absorbance will be measured at 600 nm and hence will be written A_{600}.

The number of photons scattered is proportional to the mass of the cells in a sample (except for very concentrated cultures, as discussed in the next paragraph), or, for particular growth conditions, to the cell concentration. However, the geometry of each spectrophotometer determines how much scattered light falls onto a photodetector. Therefore, a calibration curve that relates cell concentration and absorbance must be made for each spectrophotometer. This is done by measuring the absorbance of suspensions of cells at different concentrations and determining the cell concentration of each suspension by plating on agar and measuring the number of colonies formed (to be described shortly). It is important to note that the absorbance is a measure of cell mass rather than cell number. Cell size varies with growth phase, so it is best to calibrate the spectrophotometer with exponentially growing cells because such cells are used in most experiments. Cell size also varies from one growth medium to the next, decreasing as the medium becomes poorer, so a calibration curve is needed for each growth medium, and often for each bacterial strain. In this course, a single calibration curve will be used for all experiments due to time constraints. Differences in the cell sizes in different cultures have been adjusted so that they will correspond to your calibration curve.

If the cell density is too high, a photon may be deflected away from the photodetector by one cell and then back again by a second cell. This effect causes the absorbance to be lower than if multiple scattering were not occurring; it becomes important at values of A_{600} above about 0.7. Thus, when the concentration of a dense culture is to be determined by spectrophotometry, the culture is diluted prior to reading the absorbance, and the cell concentration is obtained by multiplying the measured value by the dilution factor.

Wavelengths other than 600 nm used in this course can be employed in determining cell density and, in fact, the sensitivity increases as the wavelength decreases. Wavelengths as low as 400 nm may be used, but not with all rich media. These media usually absorb short-wavelength light significantly and the absorption complicates the measurements. As rich media are used almost exclusively in this course, a wavelength of 600 nm is preferable to shorter wavelengths.

COUNTING BACTERIA BY PLATING

One bacterium can multiply until a single visible colony forms. This is the basis of counting bacteria by plating because a count of the number of colonies produced by a particular volume of a culture indicates the number of viable bacteria in the culture. In either an exponentially growing culture or a culture that has reached the stationary phase within a few hours, all bacteria can form colonies. Thus, the colony count equals the number of bacteria. To obtain a reliable count by plating, it is essential that single colonies are not formed by two or more nearby bacteria. Accordingly, the number of bacteria placed on an agar surface should not exceed a few hundred and the bacteria should be spread evenly. Since the cell concentration of bacterial cultures or cell suspensions used in laboratory experiments is usually more than 10^7 cells/ml, the culture must be diluted prior to plating. The standard procedure is to make a series of sequential 10-100-fold dilutions until the cell concentration is a few thousand cells per milliliter. Then, a 0.1-ml sample is spread on each agar surface; this operation (dilution and spreading) is called *plating*. To avoid errors that might arise by transferring very small volumes, a volume of 0.05–0.5 ml is usually transferred. In this course, we will use 10- and 100-fold dilutions prepared, respectively, by adding 0.5 ml of bacteria to 4.5 ml of sterile diluent and 0.05 ml of bacteria to 4.95 ml of sterile diluent. (Many labs use 1 to 9 and 0.1 to 9.9.) Note that if 5.0 ml of diluent are used instead of 4.95 ml, the error will be small, so often 5 ml is the value used for the 100-fold dilution. Each dilution must be done with a separate pipette; otherwise cells remaining in an earlier dilution may be carried over to a later-dilution tube. The major cause of dilution errors is the transfer of liquid on the outside of the pipette. This could be avoided by wiping the pipette, but is usually not possible because sterility would be lost. Therefore, to minimize errors, it is best to submerge the tip as little below the surface of the liquid as possible, touch the side of the tube to remove any adhering droplets, and then blow out the liquid into the diluent. The necessity to blow out the liquid is the reason that serological (blow-out) pipettes are used rather than analytical (to-deliver) pipettes.

When the final aliquot of 0.1 ml of diluted bacteria is placed on the agar, the droplet is spread over the surface with a glass spreader (Figure 2-4). The spreader is sterilized by dipping it in ethanol, shaking off the excess liquid, and igniting the remaining alcohol in the flame from a Bunsen burner. The spreader is cooled by touching the agar surface and then used to spread the droplet uniformly over the surface. If liquid remains on the agar, bacteria will drift through the liquid and, after cell division, two colonies might form from one cell initially deposited. Plates prepared

Figure 2-4 Isolation of single colonies by spreading liquid aliquots of cell cultures.

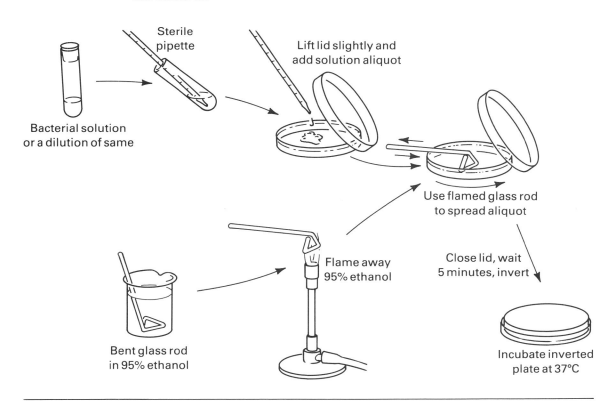

one day in advance usually absorb the liquid rapidly; absorption is accelerated by spreading the droplet as thinly as possible. After the surface of the plate is dry, the plate is placed in an incubator, typically set at 37°C. As the plate warms, liquid is exuded from the uppermost surface of the agar. Thus, it is advisable to invert the plates in the incubator to avoid puddles on the surface of the agar because bacteria will drift through the liquid. This also prevents droplets, which might condense on the inside surface of the top of the plate, from falling on the agar.

If colonies cannot be counted within 48 hours, the plates should be put upside down in a refrigerator to stop cell growth and to avoid growth of molds, which are common sources of contamination.

Chapter 3
Molecular Cloning Using Lambda Phage Vectors

E. COLI PHAGE λ

E. coli phage λ has been widely studied and has played an important role in the development of molecular genetics. Owing to the wealth of information about the biology of this phage, λ has become a common carrier DNA molecule (a cloning vector) in genetic engineering.

The DNA of phage λ is a linear double-stranded molecule, containing about 49,000 base pairs and terminated at the 5′ ends with single-stranded segments containing 12 nucleotides. These segments are complementary and can hydrogen-bond to form a double-stranded segment called *cos*, thereby producing a circular molecule.

The infectious cycle of λ begins by adsorption of a phage to a receptor protein on the *E. coli* cell wall. Following adsorption, the phage DNA is injected into the bacterium and the single-stranded termini anneal via hydrogen bonding to form a circular molecule; shortly afterwards, DNA ligase joins the strands and a covalent circular molecule is produced. Transcription then occurs and many phage and bacterial enzymes work together to replicate the circular molecules. After several rounds of replication, a second mode of replication ensues—the *rolling circle* mode—and long concatemers of λ DNA are formed. The *cos* sites on the concatemers are cut by a phage-encoded protein, and unit-sized DNA molecules result which are packaged in newly synthesized phage coats. Eventually, about 70 phage particles are formed and the cell lyses, releasing the particles to the environment. This life cycle—starting with adsorption and terminating with cell lysis—is called the *lytic cycle*.

In certain conditions, λ can engage in an alternative life cycle. The injected λ DNA directs the synthesis of a phage-encoded repressor protein that prevents production of particular mRNA molecules needed for phage production and lysis. Phage-encoded enzymes catalyze the integration of λ DNA into the bacterial chromosome at a particular base sequence called the λ *attachment site*. The bacterium does not lyse; rather the integrated λ DNA molecule, the prophage, is replicated passively as part of the bacterial chromosome (Figure 3-1). A cell containing a prophage is the lysogen. At a later time (which may be many cell generations), the λ repressor protein may be inactivated and a lytic cycle ensues, in which the prophage is transcribed, replicated, and cut from the chromosome. This process is called *induction*; it occurs spontaneously at frequencies that depend on

Figure 3-1 The life cycle of a lysogenic bacterial virus. We see that, after its chromosome enters a host cell, it sometimes immediately multiplies like a lytic virus and at other times becomes transformed into prophage. The lytic phase of its life cycle is identical to the complete life cycle of a lytic (nonlysogenic) virus. Lytic bacterial viruses are so called because their multiplication results in the rupture (lysis) of the bacteria.

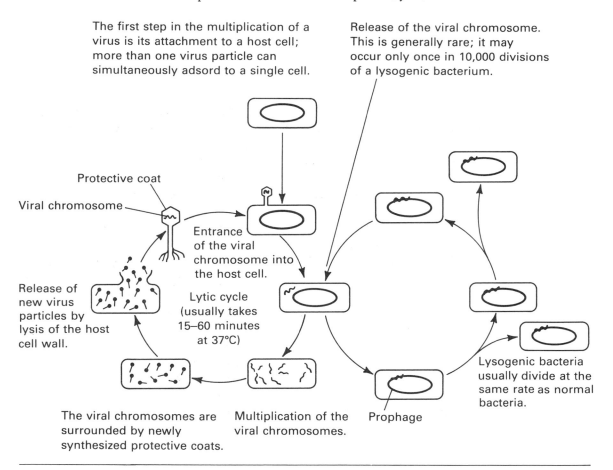

The first step in the multiplication of a virus is its attachment to a host cell; more than one virus particle can simultaneously adsorb to a single cell.

Release of the viral chromosome. This is generally rare; it may occur only once in 10,000 divisions of a lysogenic bacterium.

Protective coat

Viral chromosome

Entrance of the viral chromosome into the host cell.

Lytic cycle (usually takes 15–60 minutes at 37°C)

Release of new virus particles by lysis of the host cell wall.

The viral chromosomes are surrounded by newly synthesized protective coats.

Multiplication of the viral chromosomes.

Prophage

Lysogenic bacteria usually divide at the same rate as normal bacteria.

the growth state of the cells in a culture, but may be induced by agents that damage DNA, such as irradiation with ultraviolet light or certain antibiotics. Inactivation of the repressor is facilitated in the laboratory by using a mutant phage that makes a thermolabile repressor protein. The mutation used most frequently is called *c*I857; the repressor protein containing this mutation is active at temperatures below 32°C but is rapidly inactivated above 38°C. Thus, simply by heating a lysogenic cell above 38°C (usually 42°C is used), phage are produced 45 minutes later. Since more phage are produced by a cell growing at 37°C than at 42°C, the usual procedure is to heat a culture to 42°C for 20 minutes to complete inactivation of the repressor protein and then lower the temperature to 37°C for the remainder of the life cycle. This procedure for phage production will be used in this course.

When preparing phage, it is desirable to obtain as many particles as possible from a culture. Usually, about 70 phage λ are produced per cell, but this number can be increased to nearly 1000 by using a phage mutation called *S*7. This mutation delays lysis, but the means by which it increases the number of phage produced is much more complicated than a simple delay of lysis and is poorly understood. After a lysogen containing an *S*7 prophage is induced, the culture is allowed to grow for several hours; the bacteria are then concentrated by low-speed centrifugation. Next, chloroform is added which causes the bacteria to lyse and release the phage. A second low-speed spin removes cell debris and some bacterial DNA, leaving a phage suspension containing 10^{11}–10^{12} phage/ml.

COUNTING PHAGE BY PLATING

Phage can be counted by a modification of the plating procedure used for counting bacteria. About 10^8 bacteria are mixed with melted agar and phage particles, and this mixture is then poured onto a solid agar layer; the liquid agar cools and hardens, forming a layer about 1 mm thick. The bacteria in the thin agar layer grow for 3–5 generations and produce 10^8 microcolonies. Because the colonies are in contact, such growth produces a confluent layer of bacteria, which is visibly turbid. If a phage particle is present in the thin agar layer, it can grow in one of the bacteria initially added and produce progeny phage that can infect many nearby bacteria. Many cycles of phage growth can occur, producing 10^8–10^9 phage particles in a region no more than 1 mm across. Since the bacteria are lysed in this region, the phage particle initially present will produce a clear zone in the turbid layer of bacteria. This clear region is called a *plaque*. Since one phage produces one plaque, phage can be counted by this procedure.

Several details about plating phage λ improve the efficiency of counting the particles. First, the adsorption site on the bacterium for λ contains a component of the maltose fermentation system. Growth of the bacteria in a medium containing maltose induces the maltose operon and consequently an increased amount of the receptor protein is made. Use of maltose-grown *E. coli* accelerates adsorption of the phage, and thereby increases the size of the plaques. The number of plaques formed also increases because some plaques that are exceedingly small (owing to random delayed adsorption) become more easily visible. The presence of glucose in the growth medium suppresses formation of these receptors, so it is important to use a growth medium that is free of glucose. Second, the Mg^{2+} ion is essential for the stability of phage λ. Thus, this ion, usually in the form of magnesium chloride ($MgCl_2$) or magnesium sulfate ($MgSO_4$), is added to all solutions used for diluting and storing the phage. Third, the soft agar used in plating is fairly hot. Agar solutions have the interesting property that the melting temperature is near $100°C$ and the solidification temperature is about $43°C$. Thus, to liquefy the agar, solid agar must be heated to a temperature that would kill both bacteria and phage; bacteria and phages cannot tolerate temperatures much above $55°C$ for more than a few seconds. Thus, the procedure used is to melt the agar at high temperature and then cool it to $48-53°C$ before mixing it with phage and bacteria. The mixture is then immediately poured on a plate where it cools and hardens. It is important that the soft agar does not cool prematurely before pouring it on a plate because if it hardens too rapidly—before it has spread uniformly over the surface of the plate—the turbid bacterial background that forms will be granular rather than uniformly turbid. It is exceedingly difficult to recognize small plaques in such a granular layer, so plaque counts may be too low.

The *S7* mutation mentioned in the preceding section requires a few comments. This mutation prevents lysis and hence plaque formation. However, counting *S7* phage is possible because *S7* is a suppressible mutation. When growing in bacteria lacking a suppressor, the defective phenotype is expressed and lysis does not occur; however, if the bacteria have a suppressor—specifically, the *supF* (*sup*III) suppressor—lysis occurs and plaques can form. Thus, *supF* hosts are used for counting the phage by plaque formation, and *sup⁻* hosts are only used when large quantities of phage are to be grown.

THE *nrdA* AND *nrdB* GENES OF *E. COLI*

In the *E. coli* pathway for DNA replication deoxyribonucleoside diphosphates are formed from ribonucleoside diphosphates by the enzyme ribonucleotide reductase. This enzyme is composed of four subunits—two B1 chains and two B2 chains—that are formed by the *nrdA* and *nrdB* genes, respectively. Since there is no other pathway for production of deoxynucleotides, absence of these genes is lethal, and the mutations can be maintained in a viable cell only if they are conditional mutations. In this course, two different Nrd⁻ strains will be used; each carries the same mutation in the *nrdB* gene that decreases the B2 subunit activity but allows sufficient activity for growth of the bacterium. The presence of these mutations can be shown by observing sensitivity to hydroxyurea, a specific inhibitor of ribonucleotide reductase. A cell containing the *nrdB* mutation is capable of producing only 10% of the wild-type gene activity and is therefore inhibited by a much lower concentration of hydroxyurea than is a wild-type cell. Thus, the addition of a low concentration of hydroxyurea to agar enables you to distinguish a mutant cell from a cell containing a wild-type allele; furthermore, if a wild-type allele is introduced into a mutant cell, the cell will gain the ability to grow on such agar.

THE PHAGE VARIANT, λd*nrd*⁺

In the cloning experiments performed in this course, transfer of the *E. coli* ribonucleotide reductase genes, *nrdA* and *nrdB*, will be carried out. The initial source of these genes will be a phage λ variant, λd*nrd*⁺, which carries wild-type alleles of both genes [Collins et al. (1978); Eriksson et al. (1977)]. In this phage variant, the *E. coli* nrd genes replace several λ genes needed for phage production; the particle is defective (hence, the d in its designation d*nrd*) and the phage particle can be replicated only if the necessary genes are provided from another source. The phage DNA will be carried as a prophage in an *E. coli* strain, JF413, which also contains a normal λ prophage. When this lysogen is induced, the normal phage, which is called a *helper*, supplies the missing genes so both the helper phage and λd*nrd* phage are produced. Both prophages contain the mutations *c*I857 and *S*7, so induction is facilitated and large numbers of phage can be produced. A lysogen containing two prophages is frequently converted to a single lysogen by an intramolecular recombination process. To be sure that λd*nrd* is maintained in the lysogen, the strain JF413 also contains a mutation in the chromosomal *nrdA* gene. Thus, the bacterium only has the Nrd⁺ phenotype if the λd*nrd*⁺ prophage is present.

In the gene transfer experiments, DNA from λ and λd*nrd*⁺ will be cleaved by restriction enzymes, which cleave the DNA at defined sites, and the products of the reaction will be separated by electrophoresis through gels of agarose. Restriction maps of the λ helper phage and λd*nrd*⁺ are given in Figure 3-2 (see p. 22), which will be referred to again later.

GENERAL USE OF λ AS A CLONING VECTOR

Phage λ is a particularly valuable cloning vector for four reasons: (1) by choosing the appropriate λ vector, DNA segments of particular sizes can be cloned from a mixture of DNA fragments; (2) the recombinant DNA can be efficiently introduced into host cells after packaging the DNA *in vitro*, into λ phage particles that are capable of infection; (3) the phage containing inserted genes can be stored in a refrigerator for years without loss of viability or damage to the DNA; and (4) the DNA that is cloned can be obtained from the phage with great ease and at high concentration.

Figure 3-2 Map of the restriction endonuclease sites of λ helper and λd*nrd*. (a) *Eco*RI sites of λ helper (b) *Eco*RI sites of λd*nrd* and an expanded restriction enzyme cleavage map of the 10.94 Kbp fragment containing the *nrdB* gene. All sizes are in kilobase pairs (Kbp). The *Eco*RI fragments are designated by capital letters in order of their size, beginning with the largest which is fragment A. The three *Bgl*II sites are 30, 120, and 400 base pairs to the right of the *Bam*HI site in the 10.94 Kbp *Eco*RI fragment.

☐ λ DNA, ▨ *E. coli* DNA

E, *Eco*RI site; Bg, *Bgl*II site; Ba, *Bam*HI site; S, *Sst*I site.

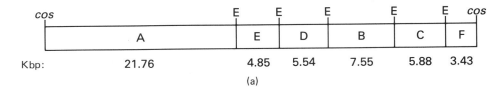

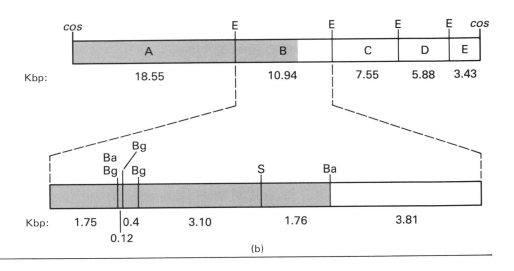

To simplify the use of phage λ as a cloning vector, λ DNA has been modified slightly to make use of a particular requirement for packaging. Phage λ can package DNA molecules ranging in size from 38.5–52.0 Kbp, that is, 78–105% of the size of the DNA normally present in wild-type particles. To produce one class of useful cloning vectors, λ DNA has been modified so that it contains two *Eco*RI restriction sites; therefore, restriction cleavage of the DNA yields three fragments. The central fragment contains genes that are needed for lysogenization but not for the lytic cycle. If the two terminal fragments are allowed to combine and introduced into *E. coli*, replication and transcription of phage DNA can occur, but phage progeny are not formed because the two terminal fragments do not form a molecule large enough to be packaged. However, if the terminal fragments are mixed with foreign DNA fragments and then the mixture is annealed and ligated to covalently link the two ends, some of the many types of molecules formed will contain both the terminal fragments and a piece of inserted foreign DNA, and will be capable of being packaged. Many different λ cloning vectors have been prepared which vary according to the number and position of the restriction sites and the particular restriction enzymes to which they are sensitive. With these variants, DNA fragments ranging in size from 1–25 kb can be cloned, although in practice 5 kb is usually the lower limit when λ DNA is used as a cloning vector [Blattner et al. (1977); Leder, Tiemeier, and Enquist (1977)].

When ligating a mixture of restriction fragments, many types of ligation products form; for example, a phage may result that has the central region that you wish to replace, the central region plus an inserted fragment, or possibly more than one inserted fragment. Thus, some selection is needed to obtain the desired phage. To eliminate those phages containing the central region, λ vectors having in this region genetic markers that make such a phage recognizable are utilized. These markers either prevent plaque formation when certain host bacteria are used or produce plaques that are visibly distinguishable and hence can be ignored. Alternatively, techniques may be used to detect inserted DNA. Some of the more specialized techniques are described in Maniatis et al. (1982). In some cases, a cleavage site is used that is within a recognizable gene—for example, the *c*I repressor gene—and insertion of a foreign fragment is detected by loss of activity of the gene. That is, a clear plaque is formed instead of a wild-type turbid plaque. A particularly useful λ variant is Charon 16A, in which the *E. coli* β-galactosidase gene has been inserted into the nonessential region of λ. Cells infected with this phage form blue plaques on agar plates containing an indicator dye. The *E. coli* gene also contains a restriction site and if DNA is inserted into this site, the recombinant phage that results will produce a white plaque because the enzyme β-galactosidase is not made.

Formation of a recombinant λ DNA molecule is not the end of the cloning process because in order to make replicas of the gene, the DNA must be packaged in a phage coat. This packaging is usually done by transfection. (*Transfection* is the uptake of viral DNA—without a viral coat—into a recipient cell.) However, it is possible to package λ DNA *in vitro* by using commercially available kits. The *in vitro* procedure involves extra steps and not all the DNA is packaged into viable particles. The greater efficiency of plaque formation from phage particles rather than from free DNA yields 20 times more plaques per weight of recombinant DNA. When a particular DNA segment is difficult to clone, this increased factor often makes the difference between success and failure.

Chapter 4
Molecular Cloning Using Plasmid Vectors

INTRODUCTION

DNA may be cloned in both phages and in plasmids. Plasmid vectors are usually much smaller than phage λ DNA molecules, and insertion of large fragments sometimes causes instability. Thus, plasmids are generally used for fragments containing less than 5000 base pairs. Plasmids are usually simpler to use than phage λ DNA molecules for three reasons: (1) plasmid DNA frequently contains only a single restriction site for a particular enzyme; (2) the ratio of inserted DNA to plasmid DNA is fairly high, making it possible to check that insertion has occurred simply by noting an increase in the size of the DNA; and (3) the supercoiling of plasmid DNA simplifies the isolation and manipulation of the DNA. However, the choice of a vector is usually determined by the size of the fragment to be cloned, the desired stability of the cloned fragment, and the purpose of cloning. Cloning in plasmids is reviewed by Bernard and Helinski (1980), Crosa and Falkow (1981), Old and Primrose (1982), and Wu (1979).

PROPERTIES OF PLASMIDS

Plasmids are extrachromosomal DNA molecules that replicate and are carried from one generation to the next. Plasmids are found in many bacterial species and often confer to the host cell phenotypic characteristics such as resistance to antibiotics and heavy metals, ability to produce particular proteins, and mating ability. Some, but not all, plasmids are transmissible from one bacterium (a donor) to another (a recipient) by a process known as *conjugation*—the pairing of two bacteria during which genetic information is transferred.

Replication of a plasmid requires that the plasmid have a replication origin and that replication enzymes are available. The enzymes are usually provided by the host cell, but the origin of replication is activated by plasmid-encoded proteins. Plasmid replication is controlled by a plasmid-encoded repressor and the interaction of this protein with sites on the plasmid DNA determines the number of copies of the plasmid per cell. If 1–2 copies are present per cell, the plasmid is a low-copy-number plasmid; more than 500 copies of a high-copy-number plasmid may be present in a host cell. Replication of plasmid DNA is usually coupled to replication of the host chromosome; if host DNA replication is inhibited, plasmid DNA replication is also inhibited. A plasmid having this characteristic is called a *stringent* plasmid. In *relaxed* plasmids, replication of plasmid DNA continues unabated when initiation of replication of the host chromosome is inhibited by an antibiotic such as chloramphenicol, which blocks protein synthesis. Relaxed plasmids are exceedingly valuable experimentally because they can be prepared in very large amounts merely by adding such an antibiotic to the culture medium. This process of increasing the number of copies (up to several thousand copies per cell) is called *amplification*.

PROPERTIES OF PLASMIDS THAT ARE USED AS CLONING VECTORS

Any plasmid can be used as a cloning vector, but a plasmid with the following characteristics is usually used.

1. It is relaxed so that large amounts of the inserted DNA can be produced.

2. It possesses distinguishable genetic markers that allow selection, usually on agar plates, of cells containing the plasmid and identification of plasmids containing inserted DNA segments.

3. It has several cloning sites—namely, specific sequences that can be cleaved by different restriction enzymes. Plasmids having cleavage sites in identifiable genes are the most useful ones.

4. It has a low molecular weight (less than 10^7). This feature facilitates the separation of plasmid DNA from chromosomal DNA and increases both the copy number and the stability of the plasmid.

Plasmids are frequently designed for particular purposes. When they are to be used for the production of specific proteins, cloning sites are usually located adjacent to regulatory elements—for example, promoters or operators—that enable one to control transcription and to make large quantities of mRNA when needed.

A plasmid commonly used for cloning and the one used in this course is pBR325 [Figure 4-1; Bolivar (1978)]. This plasmid, which was created by genetic engineering techniques, has the following features:

1. Like all plasmids, it is a covalently closed supercoiled DNA molecule.

2. It has a molecular weight of 4×10^6 (5995 base pairs).

3. It contains the replication origin of ColE1 and therefore is relaxed.

4. It contains three easily identifiable genetic markers, namely, resistance to the antibiotics ampicillin (*amp*), tetracycline (*tet*), and chloramphenicol (*cap*).

Figure 4-1 Genetic and partial restriction enzyme cleavage map of plasmid pBR325. The plasmid pBR325, a derivative of pBR322, has genes encoding resistance factors to three antibiotics and single sites of cleavage by the following restriction enzymes (the sites are given in parentheses with the *Eco*RI site as a point of reference): *Eco*RI (0/5995), *Hin*dIII (1248), *Bam*HI (1594), *Sal*I (1869), and *Pst*I (4831). The arrows show the direction of transcription of the antibiotic resistance genes. The plasmid is 5995 base pairs, as shown by the inner scale, and has a molecular weight of about 4×10^6.

amp: ampicillin resistance gene, also designated *bla* (β-lactamase)

tet: tetracycline resistance gene

cap: chloramphenicol resistance gene

ori: origin of replication

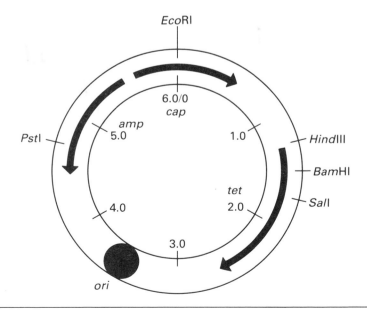

---

5. It contains one restriction site for each of the enzymes *Bam*HI, *Bgl*I, *Eco*RI, *Pst*I, and *Sal*I.

6. Each of the antibiotic resistance genes contains at least one of these sites; thus, insertion of a foreign DNA molecule into any of these sites causes loss of resistance to that antibiotic. For example, insertion in the *Eco*RI site would cause the Cap^r phenotype to be lost, but resistance to the other antibiotics would remain.

Plasmid pBR325 will be used in this course as a recipient for the *nrd*$^+$ genes, which will be taken from λd*nrd*$^+$. These genes will be inserted into the *Eco*RI site, producing a plasmid having the genotype *amp*$^+$*tet*$^+$*cap*$^-$. However, since the *nrd*$^+$ genes confer resistance to hydroxyurea (Hyu), a bacterium carrying the recombinant plasmid will have the phenotype Amp^{r}Tet^{r}Cap^{s}Hyu^r.

ISOLATION OF PLASMID DNA

Plasmid DNA is isolated from bacteria by a procedure in which cells are lysed and the chromosomal DNA, free of plasmid DNA, is precipitated. The chromosome is then removed by centrifugation and the plasmid DNA remains in the supernatant. In order to increase the yield of plasmid DNA per volume of culture, the plasmid is usually amplified by addition to the growth medium of an antibiotic that inhibits protein synthesis. Amplification not only improves the yield but increases the purity by reducing the relative amount of protein and RNA in the sample. (In the experiments in this manual, the antibiotic spectinomycin will be used for amplification.) The culture is allowed to grow overnight in the presence of the antibiotic. Then, the cells are harvested by centrifugation and treated with sodium dodecyl sulfate (SDS), an ionic detergent that lyses the cells. Addition of salt, especially potassium, to the lysate causes a chromosomal DNA-protein-SDS complex to precipitate; removal of this precipitate by centrifugation leaves a clear supernatant, known as a *cleared lysate*, which contains most of the plasmid DNA, most of the cellular RNA, and a tiny amount of broken chromosomal DNA.

The enzymatic reactions used in the cloning process are carried out most efficiently if the DNA concentration is known. The simplest way to measure the DNA concentration is by spectrophotometry. Generally, the RNA does not interfere significantly with any of the enzymatic or annealing reactions; but, since the absorption spectra of DNA and RNA are similar, the great excess of RNA interferes with the spectrophotometric determination of the DNA concentration. A variety of procedures can be used to remove RNA. The simplest is treatment of the cell extract with T1 RNase, an enzyme that hydrolyzes RNA to oligonucleotides. In addition to the removal of RNA, protein must also be removed from the solution because it absorbs ultraviolet light and will interfere with the spectrophotometric readings. A standard procedure for removing protein is to shake the solution with phenol, a reagent that forms a distinct phase with water and causes some of the protein to precipitate and some to enter the phenol phase. Consequently, shaking with phenol removes the RNase used to solubilize RNA. The RNase treatment leaves the RNA fragments in the solution. If excess protein has been removed with phenol, 80–90% of these RNA fragments can be eliminated by addition of isopropanol, a substance that precipitates DNA quantitatively but leaves most of the RNA in solution. Residual protein also remains in the isopropanol solution.

For some purposes it is advisable to remove the small amount of contaminating chromosomal DNA. This would be done, for example, if physical studies were to be performed on intact plasmid DNA. For cloning experiments this is not necessary as long as effective procedures are available for screening for bacteria that contain a recombinant plasmid.

The first recombinant DNAs made in this class will be plasmids containing DNA segments from λd*nrd*. To accomplish this, both the plasmid and the λ DNA will be cleared with a restriction endonuclease and the various fragments will be rejoined, randomly, to produce a variety of recombinant DNAs.

JOINING OF RESTRICTION FRAGMENTS

Most restriction endonucleases make two nonapposed breaks in a short inverted-repeat sequence and thereby generate short complementary single-stranded termini. For example, *Eco*RI forms the sequence AATT at the 5′ end (Figure 4-2). Since all fragments produced by this enzyme have the same termini, fragments can be joined by annealing two single strands. Joining will occur at random in the sense that any fragment has the same probability of joining to any other fragment. Furthermore, in a mixture of fragments, two competing reactions will always occur: intermolecular joining between two fragments (Figure 4-3) and intramolecular joining between the two ends of the same fragment, generating a circular mole-

Figure 4-2 Cleavage of plasmid pBR322 with *Eco*RI to produce a linear DNA molecule with two cohesive ends consisting of the single-stranded DNA sequence AATT.

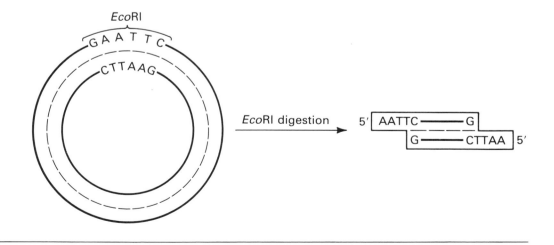

Figure 4-3 Covalent joining of two DNA molecules with complementary cohesive ends to produce a linear recombinant DNA molecule. DNA ligase catalyzes the covalent bonding after the complementary ends have annealed.

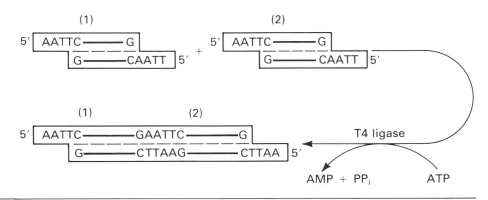

cule (Figure 4-4). A fairly complete discussion of the theory of joining and the factors influencing the relative rates of inter- and intramolecular joining can be found in Dugaiczyk, Boyer, and Goodman (1975).

Occasionally, it is necessary to prevent intramolecular joining. This can be done by treating either the donor molecule or the recipient molecule (but not both) with the enzyme alkaline phosphatase. This enzyme removes the 5′ phosphates from both ends of a fragment. This removal does not prevent annealing of the ends but it does eliminate ligation and the consequent stabilization of the molecule to heating. Usually, after recombinant DNA molecules are formed, they are treated with DNA ligase to covalently seal the single-strand breaks and prevent remelting of the joint (annealed region). The ligation reaction requires a free 3′OH and a

Figure 4-4 Covalent joining of the ends of a single DNA molecule with complementary cohesive ends to produce a circular DNA molecule.

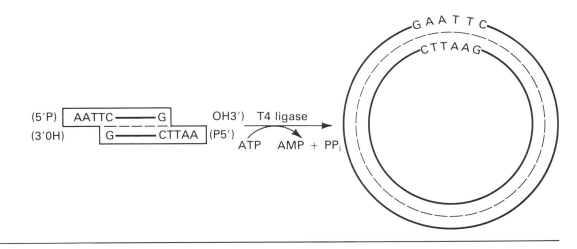

free 5' phosphate. If one of the fragments to be joined is treated with alkaline phosphatase and then circularized, no free 5' phosphates are present, ligation cannot occur, and the molecule can be remelted. However, if intermolecular joining occurs, one 5' phosphate is present in each joint so that ligase can seal *one* strand in each joint; such sealing is sufficient to prevent remelting. Heating such a mixture after ligation and before transformation linearizes plasmid molecules lacking inserted DNA and thus prohibits them from transforming cells. This procedure is not carried out in this course because the screening technique is sufficient to distinguish nonrecombinant plasmids from recombinant plasmids.

The conditions for joining and the stability of the molecules are affected by the DNA concentration (the rate increases with concentration), the size of the fragments, the presence or absence of complementary single-stranded ends (some nucleases form blunt ends and in these cases special techniques are needed for joining), the G + C content and length of the complementary single strands, the reaction temperature, and the ionic strength. The most critical experimental parameter is the reaction temperature because the melting temperature of the annealed single strands must be balanced against the optimal temperature (37°C) required for the ligation reaction. The *melting temperature T_m* is the temperature at which half of the base pairs are disrupted. If the G + C content of the single strands is low, the melting temperature may be very low. For example, the melting temperature of the *Eco*RI termini (AATT) is 4–5°C. However, 12°C is a common temperature used for sealing *Eco*RI joints. At this temperature, the joints can persist long enough (actually they are continually breaking and reforming) for ligation to occur. The ionic strength must also be balanced against the temperature because as the ionic strength is decreased, the melting temperature also decreases. On the other hand, if the ionic strength is too high, the single-stranded termini can self-anneal, forming a hairpinlike structure, and the enzyme works very slowly. The proper balance is achieved by using 10 mM and 50 mM for the divalent and monovalent ions, respectively.

In any ligation reaction in which several fragments of DNA are being joined, both intramolecular and intermolecular reactions will occur. In a particular experiment one or the other reaction is usually desired. The ratio of the two kinds of joining events can be predetermined by selection of experimental conditions. In Appendix E a simple theory is presented that enables this choice to be made.

The general scheme for cloning the *nrd* genes from λd*nrd*$^+$ into pBR325 is illustrated in Figure 4-5. In this procedure, both plasmid and phage DNA are treated with the same restriction endonuclease and all fragments are mixed. This procedure, which is called *shotgun cloning*, requires a screening procedure to find the particular recombinant DNA molecule of interest.

Four classes of DNA molecule form in the annealing mixture: plasmid DNA without inserted DNA—these have the genotype *amp*$^+$*cap*$^+$*tet*$^+$; λ fragments joined in many combinations and lacking plasmid DNA—this class does not confer any observable phenotype on the host; pBR325/λ recombinants that lack *nrd*$^+$ genes—these have the genotype *amp*$^+$*tet*$^+$*cap*$^-$ and do not confer hydroxyurea resistance (Hyu^r) on a host bacterium; and

Figure 4-5 Flow diagram of the subcloning of the *E. coli nrd* genes, from phage λd*nrd*, into plasmid pBR325. (a) both pBR325 and λd*nrd* are cleaved with *Eco*RI. (b) The fragments are annealed and covalently joined by DNA ligase. Some of these products will be plasmids containing inserted λd*nrd* DNA fragments. This step selects for cells transformed with plasmids; they will grow in an environment containing ampicillin. (c) *E. coli* cells are transformed with the mixture of joined DNAs and grown on agar plates containing ampicillin. (d) *E. coli* cells resistant to ampicillin are replica plated on agar plates containing either chloramphenicol or hydroxyurea to distinguish the various types of plasmids. (e) On the basis of their phenotype, cells harboring recombinant plasmids with the *nrd* insert (class iv) can be identified. E = *Eco*RI restriction endonuclease cleavage site.

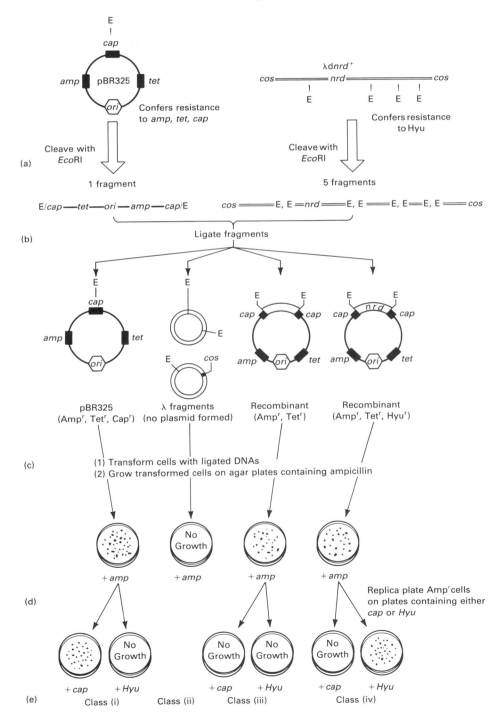

Figure 4-6 Formation of a circular recombinant DNA. Two DNA molecules with complementary cohesive ends are annealed and covalently joined to form a linear recombinant DNA which is then circularized via intramolecular ligation. Both steps occur in the same reaction vessel and are catalyzed by DNA ligase.

a pBR325/*nrd*$^+$ recombinant—a *nrd*$^-$ host containing one of these plasmids will have the phenotype Amp^{r}Tet^{r}Cap^{s}Hyu^r.

In order to subclone the *nrd*$^+$ genes from λd*nrd*$^+$ into pBR325, both intramolecular and intermolecular joining must occur: first, pBR325 and the λd*nrd*$^+$ fragments must join intermolecularly to form a recombinant product, and, second, the recombinant molecule must circularize to form a stable recombinant plasmid, as shown in Figure 4-6. Other recombinants —for example, λ/λ, plasmid/plasmid, and molecules containing three or more fragments—will also form.

TRANSFORMATION OF BACTERIA BY RECOMBINANT DNA MOLECULES

Once a recombinant DNA plasmid has formed, it is necessary to establish it as a replicating element in a bacterium in order to maintain it. This is accomplished first by altering a recipient cell, so that DNA can enter the cell (making the cells *competent*), and then by adding the recombinant DNA to a suspension of these cells. Several procedures, called *transformation* procedures, are available; in this course, the CaCl$_2$ method [see Hanahan (1983)] will be used. In this procedure, bacterial cells are swollen in a hypotonic buffer containing calcium chloride (CaCl$_2$) and briefly heated; by this protocol, plasmids can be established in as many as 0.1% of the cells, resulting in up to 10^6 transformed cells per microgram of recombinant plasmid DNA molecules.

SCREENING OF TRANSFORMED CELLS FOR RECOMBINANT PLASMIDS

Screening refers to the process of selecting, among those cells that have taken in DNA, cells that have the desired phenotype. In the experiments in this manual, the procedure begins by selecting cells in which a pBR325 plasmid is established. This selection is accomplished by plating the cells that, without the plasmid, have the Amp^s phenotype on agar containing ampicillin. Only Amp^r colonies can form and these will almost always contain pBR325 with or without an inserted fragment of DNA. The Amp^r colonies will be either Cap^s or Cap^r; if insertion has occurred in the *cap* gene, they will be Cap^s. Thus, colonies are taken from the ampicillin plate and tested for sensitivity to chloramphenicol. These Amp^{r}Cap^s colonies can also be either sensitive or resistant to hydroxyurea; if the *nrd* genes have been obtained from the λd*nrd*$^+$ DNA, the cells will be Hyu^r. Thus, to select the desired colonies the Amp^{r}Cap^s colonies are transferred to a plate containing hydroxyurea and colonies that grow are picked. The recombinant pBR325 plasmids containing an *nrd*$^+$ gene insert can be found directly by plating the transformed cells on an agar plate containing both ampicillin and hydroxyurea.

PHYSICAL ANALYSIS OF RECOMBINANT PLASMIDS

As mentioned earlier, recombinant plasmids may contain multiple copies of particular DNA segments. Whether one or more copies are present can be determined by physical analysis of plasmid DNA isolated from Amp^{r}Cap^{s}Hyu^r cultures. The plasmids are treated with restriction enzymes having sites in both the plasmid DNA and the λ phage DNA and the sizes of the fragments produced by this treatment are measured after electrophoresis in agarose gels.

Since the termini of fragments generated by a restriction enzyme are identical, the *nrd*$^+$ genes can be inserted into pBR325 in two different orientations. The orientation can also be determined by restriction analysis of the recombinant plasmids, because the restriction sites are located such that fragments of different sizes will be produced from plasmids having the *nrd*$^+$ genes in each orientation. From the restriction maps for pBR325 (Figure 4-1) and λd*nrd*$^+$ (Figure 3-2), the restriction maps for the two types of recombinant plasmids can be predicted. Whether both orientations are present is of interest because it provides information about whether the promoter for the *nrd* gene is included in the fragment. If the promoter is absent, expression of the *nrd* gene can only occur if the gene is coupled to a promoter already in the plasmid. Thus, since the plasmids are selected by their resistance to hydroxyurea, only a single orientation would normally be found, if coupling to a plasmid promoter were necessary for gene expression. However, if the *nrd*$^+$ segment contained the *nrd* promoter, plasmids having the *nrd*$^+$ genes in both orientations will be found.

Chapter 5

Molecular Cloning Using M13 Vectors

INTRODUCTION

E. coli phage M13 presents a new mode of cloning—namely, cloning in a single-stranded DNA molecule.

PHAGE M13

M13 is a filamentous phage that can only adsorb to strains of *E. coli* containing the F sex plasmid. The phage particle contains one circular single-stranded DNA molecule, called the *(+) strand*. Following infection, the single-stranded DNA molecule is converted to a double-stranded circular replicating form, RF, by synthesis of a complementary strand termed the *(−) strand*. More RF molecules are made throughout the life cycle [Kornberg (1980)] and, later in the life cycle, progeny (+) strands are made and packaged in the phage coat. Packaging of M13 differs from that of most other phages in that there seems to be no limit to the amount of DNA that can be packaged. Since the particle is filamentous, packaging of a larger circular DNA molecule merely results in formation of a longer filament. Another unusual feature of the life cycle is that infected cells do not lyse; instead, progeny phage particles are extruded from an infected cell. This

Figure 5-1 Photograph of phage M13 plaques on a lawn of *E. coli*. The colorless plaques are produced by recombinant M13mp11 that contains DNA in the polycloning site (as described in Figure 5-3 below) and the dark plaques are produced by wild-type M13mp11 that lacks a DNA insert in the polycloning site.

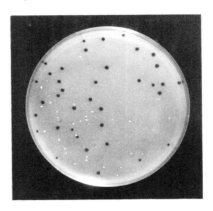

property has a significant effect on the morphology and timing of appearance of plaques, because plaques are normally the result of lysis of bacteria in a bacterial lawn. Infected cells grow more slowly than uninfected cells; thus, an M13 plaque is a region in which cells are growing more slowly than neighboring cells. The more slowly growing infected cells, in the lawn of more rapidly growing cells, produce turbid plaques (Figure 5-1). An outline of the essential features of the M13 system is given in Figure 5-2.

USES OF M13 AS A CLONING VECTOR

No simple method is available at present for joining two isolated single strands. Thus, when cloning in M13 is desired, the recipient molecule is the double-stranded RF. This molecule is isolated by the same procedure employed for isolating plasmids and its use as a cloning vector is exactly like that of a plasmid DNA molecule. That is, the RF is cleaved with a restriction enzyme, mixed with DNA fragments, and the mixture is annealed and ligated. Cells made competent by exposure to calcium chloride ($CaCl_2$) are transfected with ligated DNA and progeny phage particles are obtained. These particles are then tested for the presence of inserted DNA.

THE M13 CLONING VECTOR

Joining of any cleaved circular DNA molecule with a restriction fragment results in the formation of recombinant molecules in which the inserted

Figure 5-2 The life cycle of phage M13. After the phage enters the host cell through a pilus, the single-stranded genome is released and copied to form a double-stranded replicating form of DNA, the RF. The RF directs the synthesis of progeny, single-stranded M13 genomes which are packaged and extruded from the host cell. The uncapsulated RF DNA can be directly taken up by the cell using a hypotonic, calcium phosphate/heat-shock transformation procedure similar to that used for plasmid transformations. A change occurs in the outer membrane protein of the phage, shown in the figure as ■→●, during the extrusion process.

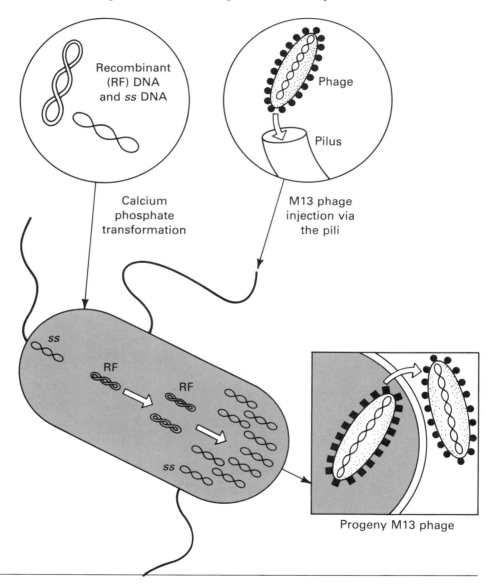

fragment can have one of two possible orientations. If the two strands of the inserted fragment are called *A* and *B*, then, in some recombinant molecules strand *A* is linked to a (+) strand and in other recombinant molecules strand *B* is linked to a (+) strand. Since only (+) strands are packaged, the

A and *B* strands of any segment can be cloned separately. This property is an important reason for using M13. For example, separately cloned single strands can be used in hybridization experiments in which one wishes to assay mRNA without interference by the noncoding strand; also, base sequencing is done more conveniently if pure single strands are available.

PRECAUTIONS IN THE USE OF M13 AS A CLONING VECTOR

As already mentioned, cells infected with M13 grow more slowly than uninfected cells. For unknown reasons, as the size of the DNA in the phage particle increases, infected cells grow even more slowly. This alteration in growth rate affects the cloning strategy because the source of recombinant phages is always a growing infected culture. If a culture is infected with DNA molecules covering a range of sizes, cells containing the larger molecules will grow more slowly than those containing the smaller DNA molecules, so there is a selection against the larger molecules. Therefore, in contrast to phage λ and plasmids, M13 vectors carrying DNA inserts cannot be propagated as a mixture—transformants must be plated directly to produce single plaques. The collection of different recombinants is called a *recombinant DNA library*. A complete library of a mammalian genome, that is, a collection of recombinants that contains at least one copy of every gene, may contain 10^6–10^7 different recombinant DNA molecules.

COMPONENTS OF THE M13 CLONING SYSTEM

The M13 cloning system has two components—the M13 phage itself and the host cell. Together these components allow insertion of foreign DNA to be detected by loss of activity (*insertional inactivation*) of a functional segment of the *E. coli lac* operon carried by the phage. A simple color assay measures the activity. The system has been designed so that the two complementary strands of a DNA sequence can be separately cloned.

The mainstay of the insertional inactivation test is a histochemical assay based on the ability of β-galactosidase, a product of the *lac* operon, to cleave certain synthetic substrates. The presence of this enzyme in active form can be monitored by the enzymatic cleavage of a colorless compound, 5-bromo-4-chloro-3-indolyl-β-D-galactoside (X*gal*), to form galactose and the deep blue substance, 5-bromo-4-chloroindigo. This test can be performed on an agar plate, as shown in Figure 5-1.

The simplest way to design a system based on inactivation of the *lac* operon would be to include the complete operon in the vector. However, this would make the vector quite large. By genetic engineering, the system is made to function with only a small part (447 nucleotides) of the sequence of the β-galactosidase gene (3063 nucleotides). This sequence encodes only the amino terminal end of the enzyme. The presence of this fragment in a functional state can be detected by an intracistronic complementation test.

Intracistronic complementation refers to the production of a functional protein by two genes, each of which contains a mutation in the same polypeptide chain. In the *lac* system, complementation can occur because the active form of β-galactosidase is a tetramer. For example, consider two defective enzyme molecules encoded in two mutant *lac* operons, one in the *E. coli* chromosome and one in the F plasmid. In the haploid state the phenotype of a cell containing either of these molecules will be Lac^-; however, in a partial diploid the two polypeptides may be able to interact directly with each other to form a functioning enzyme [Ullmann, Jacob, and Monod (1967)]. This will not always be the case since some mutant proteins either may not be able to interact or would interact in such a way that an active site is not generated.

Intracistronic complementation can also occasionally occur with polypeptide fragments, although this is not usually the case. A particularly useful example of this phenomenon serves as the basis of the test with the M13 system. The *lac* deletion mutation *lacZ*(ΔM15) or simply M15 (not to be confused with phage M13) causes synthesis of a polypeptide that lacks amino acids 11–41 (of a total of 1021 amino acids). The deletion is not in the active site of the enzyme but it does prevent tetramerization. Inactivity of the M15 polypeptide can be overcome *in vitro* by the addition of a particular cyanogen bromide fragment (amino acids 2–92) of the wild-type protein [Langley et al. (1975)]—somehow, this fragment enables the M15 polypeptides to form a tetramer. Thus, the fragment is said to complement the M15 mutation. This phenomenon is called ***alpha complementation***; the M15 protein is said to be an alpha acceptor and the cyanogen bromide fragment an alpha donor. Alpha complementation also occurs if, instead of the cyanogen bromide fragment, a particular *Hind*II fragment that contains the *lac* regulatory region (operator and promoter, but not the repressor) and 15% of the amino terminal end of the β-galactosidase gene is present [Landy et al. (1974)]. In the experiments in this course the *lac Hind*II fragment is contained in the M13 phage and the gene containing the M15 deletion is carried on an F plasmid in the bacterium.

Although this system is sufficient for cloning, it is always desirable to be able to regulate the activity of a cloned gene. This is made possible by the presence of the *lac* operator and promoter in the *Hind*II fragment carried in the M13 vector; however, because the fragment does not carry a functional *lac* repressor gene, constitutive synthesis would occur in an infected cell for the following reason. When a bacterium is infected with an M13 particle (recombinant or otherwise), about 200 copies of M13 RF quickly accumulate in every cell. In the case of an M13 carrying the *Hind*II fragment, repression of an ordinary Lac^+ cell would be overcome by this replication because only about ten molecules of repressor would be present in each cell. Consequently, in an infected cell, transcription of the *lac* structural genes is constitutive [Heyneker et al. (1976)]. In order to overcome this constitutivity, a mutation—I^q—in the *lac* repressor gene [Calos (1978)], which causes a tenfold overproduction of repressor [Müller-Hill, Crapo, and Gilbert (1968)], is put in the host cell. In the presence of this mutation enough *lac* repressor is made to compensate for the large number of M13 RF molecules [Messing et al. (1977)]. In a host cell with the *lac*I^q mutation, transcription of the *Hind*II fragment can be activated by adding an inducer of the *lac* operon; typically, the inducer isopropyl-β-D-thiogalactoside (IPTG) is used. In the color test used to detect complementation, the *lac* operon must also be induced. X*gal* is not an inducer so it is also necessary to include IPTG in the agar.

Another feature of the M13 system, required not for cloning but to satisfy certain legal requirements, is described in the following section.

BIOLOGICAL CONTAINMENT OF M13

During the period in which recombinant DNA technology was developing, many concerns were expressed about the possible consequences of recombinant DNA molecules escaping to the environment. Thus, vector-host systems were constructed that, for propagation, depend on specific conditions found in the laboratory but not in nature. An important decision was the use of *E. coli* K12 strains, which are unable to grow in the human gut.

Containment of plasmids created a special problem because they may be transferred to other bacteria that could grow in the environment. Plasmid transfer depends on the presence of conjugative transfer genes contained in both the plasmid and the host. (In hospitals, the conjugative properties of bacterial plasmids are well known since they are responsible for the spread of the antibiotic resistance genes that are encoded in many plasmids.) To avoid this problem, plasmid vectors that do not conjugate have been used exclusively. However, growth of M13 depends on several genes of F, a plasmid not used for recombinant DNA work because of its high frequency of conjugation. The presence of F in the same host as the M13 might permit the spread of recombinant DNA molecules to other organisms if the recombinant DNA molecule were to recombine with F and yield a transmissible recombinant DNA. However, the problem is not serious for two reasons. First, conjugative transfer of F plasmids is reduced 10^6 times in M13-infected cells compared to uninfected cells. Second, this value can be reduced further by introducing a mutation into F (for example, *traD*); this mutation alone reduces conjugative ability 10^5 times [Achtman, Willets, and Clark (1971)] without affecting the ability of the cell to serve as a host for M13. Thus, the transfer rate from an M13-infected cell is reduced 10^{11} times overall [Messing (1979)].

In summary, the bacterial strain used in the M13 cloning experiments has the following genotype: JM107 Δ(*lac pro*) *thi*-1 *gyr*A-96 *end*A1 *sup*E44 *rel*A1 *pro*A1 *pro*AB⁺ *lac*I^q (ΔM15).

THE M13mp SERIES

A series of M13 cloning vectors has been prepared that facilitates cloning of single-stranded DNA molecules. The series is designated M13mp*n*, in which n is an integer. These vectors possess two especially useful properties. First, they contain a modified β-galactosidase gene fragment (the *Hind*II fragment) with a genetically engineered insert that has a high concentration of unique cloning sites in a small locus [Messing and Vieira (1982)]. Such a concentration of sites is known as a *polycloning site* (see Figure 5-3). This modified gene has been described earlier. Second, the phages have been engineered in pairs in order to permit efficient cloning of both DNA strands. The second property of these vectors has not been widely discussed, so is presented here in slightly greater detail than the treatments of phage and plasmid vectors given in previous chapters.

The M13 vector can be used to provide a source of separated strands of a particular sequence because a fragment cloned in M13 RF can be obtained as a single strand in phage progeny. However, to have a source of both strands requires two separate cloning events. The particular strand of a cloned fragment that is produced by a recombinant phage is determined by the orientation of the DNA inserted into the RF because the particle contains only a (+) strand. In a population of RF containing inserts, both orientations of the inserted DNA do not occur with equal probability. One possible solution to this problem is to use a screening procedure to detect phages with each orientation. An alternative, which we use in this course, is to utilize a cloning procedure in which the orientation of insertion is directed. The latter can be accomplished if two different restriction enzymes are used.

A DNA fragment produced by two restriction enzymes (for example, *Bam*HI and *Hind*III) has two different ends. A vector cleaved by these two enzymes cannot recircularize unless another fragment having the same two termini is present; thus, any phage that results from transformation with doubly cleaved RF must contain both phage and target DNA. To produce such a recombinant, the *Bam*HI ends of the vector and of the inserted DNA must be connected and the *Hind*III ends also must be joined. The result of such a treatment is a defined orientation of insertion, as illustrated in Figure 5-4. Since the location of the *Bam*HI and the *Hind*III sites in the vector are known, the orientation of the inserted DNA fragment can be predicted.

Sometimes the target DNA is cleaved with a restriction enzyme that produces overlapping ends that are compatible for cloning into the vector even though the recognition site is different. For example, the recognition site for *Bam*HI in the M13 vectors is $\genfrac{}{}{0pt}{}{\text{GGATCC}}{\text{CCTAGG}}$ which produces molecules looking like $\genfrac{}{}{0pt}{}{\text{GATCC-G}}{\text{G-CCTAG}}$. This DNA can be ligated to a DNA cleaved with *Bgl*II whose recognition sequence is $\genfrac{}{}{0pt}{}{\text{AGATCT}}{\text{TCTAGA}}$ and which produces molecules with overhanging ends $\genfrac{}{}{0pt}{}{\text{GATCT-A}}{\text{A-TCTAG}}$. These two molecules have complementary ends and can be ligated together to form a junction that is not recognized by either *Bam*HI or *Bgl*II. We make use of this homology of overlapping ends during the cloning of pBR*nrd* fragments, produced by *Bgl*II and *Sst*I digestions, into M13mp10 and M13mp11 cleaved with *Bam*HI and *Sst*I.

Figure 5-3 Structure of a pair of M13 cloning vectors containing a poly-cloning site that is in either of two orientations. The restriction endonuclease cleavage sites which are unique to the M13mp10 and M13mp11 pair are *Eco*RI, *Sst*I, *Xma*I/*Sma*I, *Bam*HI, *Xba*I, *Sal*I/*Acc*I/*Hind*II, *Pst*I, and *Hind*III. The vectors do not have recognition sites for *Ava*II, *Bgl*I, *Bst*EII, *Kpn*I, and *Xho*I. The M13 genes are indicated by roman numerals.

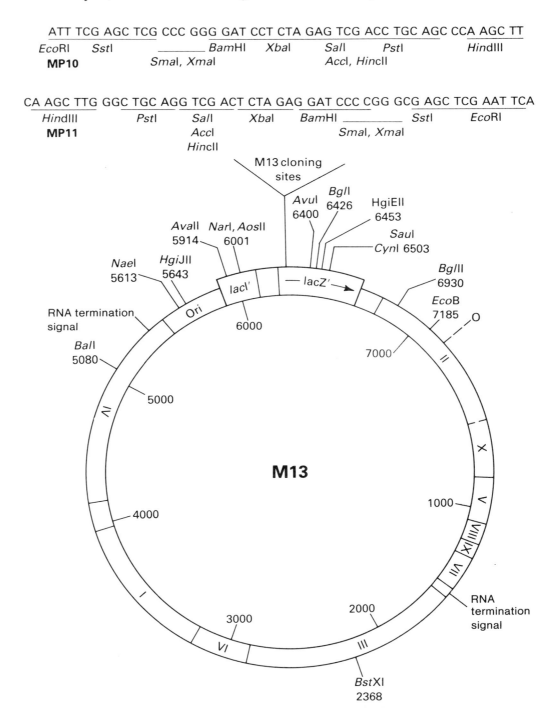

ATT TCG AGC TCG CCC GGG GAT CCT CTA GAG TCG ACC TGC AGC CCA AGC TT

| *Eco*RI | *Sst*I | | *Bam*HI | *Xba*I | *Sal*I | *Pst*I | *Hind*III |

MP10   *Sma*I, *Xma*I   *Acc*I, *Hinc*II

CA AGC TTG GGC TGC AGG TCG ACT CTA GAG GAT CCC CGG GCG AGC TCG AAT TCA

| *Hind*III | *Pst*I | *Sal*I | *Xba*I | *Bam*HI | | *Sst*I | *Eco*RI |

MP11   *Acc*I   *Sma*I, *Xma*I

                       *Hinc*II

M13 cloning sites

*Avu*I 6400   *Bgl*I 6426   *Hgi*EII 6453

*Ava*II 5914   *Nar*I, *Aos*II 6001   *Sau*I

*Cyn*I 6503

*Nae*I 5613   *Hgi*JII 5643   *lac*I'   — *lac*Z' →   *Bgl*II 6930

RNA termination signal   Ori   *Eco*B 7185   O

*Bal*I 5080   6000   7000   II

5000   M13   X

IV   V

4000   1000   VIII

I   XI/XII   VII

3000   2000   RNA termination signal

VI   III

*Bst*XI 2368

Figure 5-4 Forced cloning of DNA fragments into the polycloning sites of M13 using doubly digested target and vector DNAs. The restriction endonucleases are identified by single letters: B, *Bam*HI; E, *Eco*RI; H, *Hind*III; S, *Sal*I.

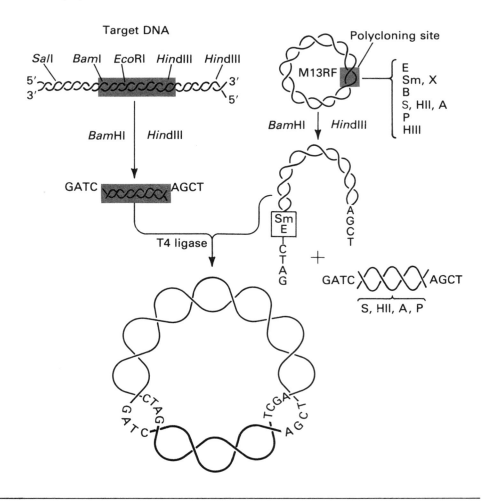

A single M13 vector treated with two restriction enzymes can only yield one of the two strands of a DNA fragment; thus, to obtain both strands a second M13 vector having a reversed order of the *Bam*HI and *Pst*I sites is needed. Such a pair of vectors containing several reversed cloning

Figure 5-5 Structure of the polycloning sites in M13mp7–M13mp11 and in the associated pUC plasmids. The position of the polycloning site in M13 is shown at the bottom; the position of the polycloning site in the pUC plasmids is shown at the top. The sequences in the amino terminal ends of the mutated β-galactosidase alpha subunits are shown along with the resulting amino acid sequences and *unique* restriction endonuclease cleavage sites.

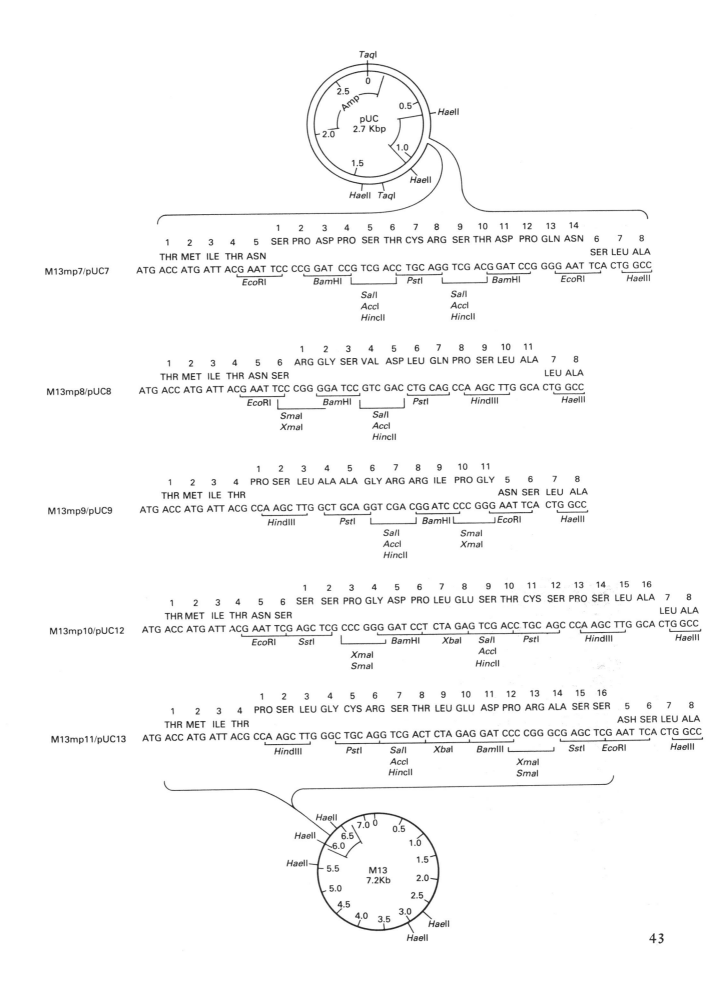

sites has been developed. The pair used in this course is M13mp10 and M13mp11 [Messing and Vieira (1982)].

A description of the different M13mp vectors in common use is given in Figure 5-5; the pertinent cloning sites and their codon positions in the *lac* DNA are indicated. Note that complementary plasmids, the pUC*n* plasmids, have been constructed to facilitate moving DNA inserts between plasmids and M13 vectors [Vieira and Messing (1982)].

DETERMINATION OF ORIENTATION OF INSERTED DNA

Two DNA molecules isolated from M13mp10 and M13mp11 cannot hybridize with one another because each is a (+) strand. However, the inserted DNA can have either polarity in the DNA of the particle so two M13 recombinants, having the same cloned DNA but in two different orientations, have complementary inserted sequences and hence can hybridize. The hybridization is restricted to the inserted DNA sequences, so the M13 sequence stays single-stranded; thus, the two interacting single-stranded recombinant M13 molecules anneal to form a structure resembling a figure-8 (Figure 5-6). The production of this structure can be used

Figure 5-6 Analysis of cloned DNAs by the C-test. DNA clones that contain complementary strands of target DNA can be hybridized to form dimers which migrate slower during electrophoresis through agarose gels than do unhybridized monomers. There may, or may not, be contaminating chromosomal DNA and RNA in the phage DNA preparations. The bands in the C-test gels are generally broader and more diffuse than those in other analytical gels because of the high concentration of salt in the hybridization solution, which is transferred to the wells of the gel.

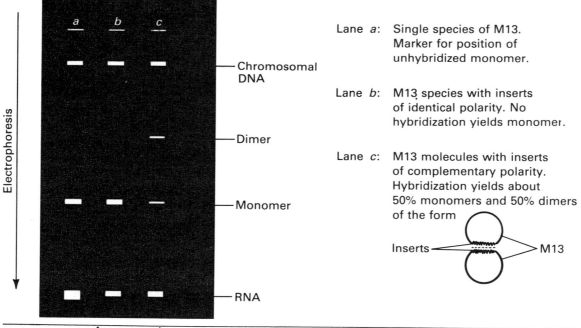

Lane *a*: Single species of M13. Marker for position of unhybridized monomer.

Lane *b*: M13 species with inserts of identical polarity. No hybridization yields monomer.

Lane *c*: M13 molecules with inserts of complementary polarity. Hybridization yields about 50% monomers and 50% dimers of the form as a simple test, known as the *C-test* [Messing (1983)] , for studying the orientation of the inserted sequence. The hybridized figure-8 has a lower electrophoretic mobility in an agarose gel than the individual parent molecules and is easily detectable. Thus, the presence of a slowly moving band when two recombinant (+) strands are annealed indicates that hybridization has occurred and that each DNA molecule contains an inserted sequence complementary to that of the other. If a single sequence has been inserted, the orientation in each phage must be in the opposite sense to that of the other phage. The C-test can also be used to determine whether particular M13 subclones are related [Howarth et al. (1981)] .

SITE-SPECIFIC MUTAGENESIS

A detailed analysis of the function of a particular gene ultimately requires determination of the nucleotide sequence. The first stage in this process—isolation of the DNA—is made possible by the recombinant DNA techniques we have described. Once the nucleotide sequence is obtained, questions about the relation between sequence and function can be answered. An important stage in answering such questions is the determination of the components of the sequence that control expression of the gene. The strategy used to analyze gene expression has always been to study mutants. With the nucleotide sequence in hand, the order of mutants having altered (usually reduced) function can be determined, and you can observe the base changes that influence regulation, initiation of transcription, termination, and so forth.

Traditionally, mutations are obtained by the random action of mutagens. Sometimes these mutations occur rarely, and only a few nucleotide changes can be accumulated after a great deal of work. Here we describe a technique that enables mutations to be created in particular sequences, efficiently and, if needed, in large numbers. The principle underlying the technique is the use of a short synthetic oligonucleotide primer to initiate replication of a single-stranded DNA molecule. The primer becomes a part of the daughter strand and thereby gives rise to a new strand having an altered sequence. In order to obtain site specificity, the primer is designed to have the nucleotide sequence—identical except for one or a few bases—of a particular region of the gene to be studied. The procedure, which is shown in Figure 5-7, follows.

A section of the known sequence of a gene to be studied is chosen. Using the technique of Gillam and Smith (1979) a short single-stranded oligonucleotide (13–30 bases) of defined sequence is enzymatically synthesized by stepwise addition of particular nucleotides—a template is not used. (Usually, the oligonucleotide is synthesized *de novo*, but occasionally one starts with an oligonucleotide obtained by restriction cleavage—followed by strand separation—of the gene being studied.) The oligonucleotide is complementary to part of the presumed regulatory sequence of the gene but contains a mutation. The oligonucleotide is then annealed to a single-stranded version of the gene being studied and used to prime

Figure 5-7 Site-specific Mutagenesis

(1) Wild-type M13 with a region (x) for a site-specific mutagenesis is annealed to a primer DNA. The primer is complementary to a region of M13 except at the site of the specific mutagenesis (■). The M13 is the (+) strand, the primer is the (−) strand. (2) The heteroduplex is taken up by a host *E. coli* cell. (3) The host cell's enzymes complete the (−) strand to form a heteroduplex RF. (4) DNA replication using the (−) strand as a template produces, via a rolling circle mechanism, many copies of the (+) strand which contains a mutation (▨) complementary to the primer mutation. Wild-type phage are produced from other (−) strands synthesized in the cell from the original (+) strand of the heteroduplex. (5) Both mutant and wild-type M13 are released from the transformed cells.

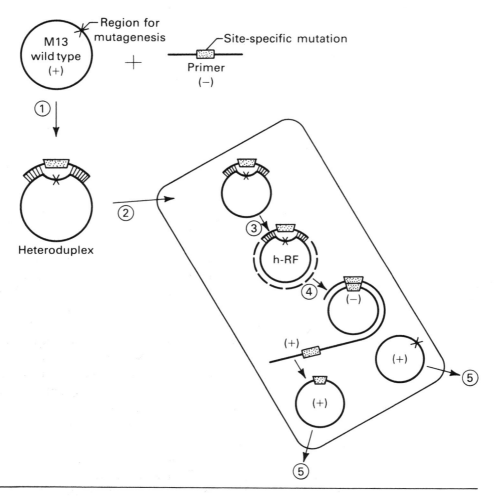

copying of the strand [Hutchison et al. (1978)]. For a gene cloned in M13 the oligonucleotide would be a primer for (−) strand synthesis and would be annealed to (+) strand DNA obtained from the phage particles. Annealing to the template strand must be done at low temperatures because the double-stranded segment is generally quite small and has low thermal stability, and the mispaired region further destabilizes the structure (known as the *primed strand*). In general, the annealing temperature depends on the length of the primer and the location of the mispaired region with respect to the ends of the double-stranded segment [Gillam and Smith (1979)]. The primed strand can be used in an *in vitro* reaction with DNA polymerase I to synthesize a double-stranded RF molecule. More commonly, competent cells are transformed with the primed strand, and synthesis of the RF occurs *in vivo*. The infected cells produce two classes of progeny phage—wild-type phage and those containing the new mutant sequence. Plating these phage yields pure clones of particles having the mutant sequence. Determination of the phenotype of the mutant phage gives the desired information about the role of particular bases in regulation.

In this manual a variation of the above procedure will be used to avoid the necessity of the time-consuming and expensive synthesis of the oligonucleotide fragment. An M13 variant containing a mutant *lac* fragment (it contains a frame shift) will be altered by the site-specific technique. The primer will be a wild-type sequence obtained by restriction enzyme cleavage of a *lac* segment cloned in a plasmid. Thus, site-specific mutagenesis will be used to create a revertant rather than a defective mutant.

Part Two
Laboratory Exercises

Period 1

Introduction
to the Course

I. INTRODUCTION

During Period 1 of this course, you will have a chance to familiarize yourself with the physical arrangement of the laboratory and the common procedures you will use in the course. The first half of the period will consist of a lecture in which the course will be outlined and in which the properties of the double-stranded DNA phage λ will be described. During the second half of the period you will isolate single colonies of the *E. coli* strains to be used in the next two periods. The strains needed are JF335, JF413, JF417, and JF427. Strain JF413 will be used to prepare λdnrd^+ and a λ phage (which we term a helper) that is needed for growth of the defective λdnrd^+ phage; JF335 will be used as an independent source of the helper phage. In Period 7, you will compare the phages in JF413 and JF335 so that the DNA containing the nrd^+ genes can be identified. The other strains will be used for detection of the two phages. JF417, which contains a *supF* mutation that allows suppression of the S7 mutation of the λ helper phage, will be used for titering the helper phage. JF427 will be used as a recipient for detecting λdnrd^+; it is $nrdB^-$ and will not grow on agar containing 0.75 mg/ml hydroxyurea unless it receives a functional nrd^+ gene. A complete list of the bacterial strains to be used in this manual is given in Table P1-1.

Since each strain has particular characteristics that will be needed, it is essential that the strains be pure. Often strains become contaminated with other bacteria or acquire mutations that would ruin an experiment. Strains are purified by streaking them on an agar surface, selecting a single colony, and testing this colony for the characteristics needed. As an exam-

Table P1-1   A complete list of bacterial strains

Strain	Genotype, Properties, and Application
JF335	λ(cI857 s7)

This strain will be used as a source of λ DNA that will be used for markers in gel electrophoresis. This strain will grow at 30°C but not at 42°C since the cI857 mutation causes induction of a lytic cycle of λ.

JF413	araD139Δ(ara leu) 7697 ΔlacX74 galU galK hsdR strA gyrA nrdB1 malB λ (b505, 519 cI857 s7 xis6) λd(b515, 519 cI857 s7 xis6 nrdA⁺B⁺)

This strain will be used as a source of λnrdA⁺B⁺ DNA. The nrdB mutation prevents growth on agar plates containing 0.75 mg/ml hydroxyurea at 30°C unless the λd(nrdA⁺B⁺) phage is present. This test will be used to verify that the λd(nrdA⁺B⁺) is still present in the strain. This strain will grow at 30°C but not at 42°C due to the cI857 mutation. The malB mutation eliminates the λ receptors on the cell surface and prevents attachment of phage λ to cell debris during growth of the phage.

JF417	supE44 supF58 lacY or Δ(lacIZY)6 galK2 galT22 metB1 trpR55 hsdS

This strain will be used to titer phage λ. The supF58 mutation suppresses the S7 amber mutation and allows λ to form plaques.

JF427	araD139 Δ(ara leu) 7697 ΔlacX74 galU galK hsdR strA gyrA nrdB1

This strain will be used as a recipient for transformation of plasmids containing the nrdA⁺B⁺ genes. The nrdB1 mutation makes this strain sensitive to 0.5 mg/ml hydroxyurea at 37°C. Transformants that obtain the nrdA⁺B⁺ genes will grow under these conditions. This strain is used since it gives a high transformation efficiency. The hsdR mutation eliminates the endonuclease R that is responsible for degradation of DNA that does not contain K-strain modified DNA. Thus, foreign DNA can be placed in this strain without its being degraded.

JF428	araD139 Δ(ara leu) 7697 ΔlacX74 galU galK hsdR strA plasmid pBR325

This strain will be used as a negative control for the testing of strain JF427 into which a pBR325 derivative containing nrd⁺ has been inserted.

JF429	araD139 Δ(ara leu) 7697 ΔlacX74 galU galK hsdR strA gyrA nrdB1 plasmid pBRnrd

This strain will be used as a positive control for the testing of strain JF427 into which a pBR325 derivative containing nrd⁺ has been inserted. Plasmid pBRnrd contains an EcoRI fragment, from λnrd⁺, that contains the nrd genes.

JM107	Δ(lac pro)thi-1 gyrA96 endA1 supE44 relA1 proA⁺B⁺ lacIq lacZ(ΔM15).

This strain is used as a host for phage M13.

ple, consider the purification of JF413, the strain that contains both λdnrd⁺ and the helper phage. Bacteria from a slant culture are streaked on RA/Hyu agar, a simple and inexpensive agar (RA) containing hydroxyurea (Hyu). Only a cell containing a wild-type nrd allele can form a colony. JF413 has a mutant nrdA gene in its chromosome but the nrdA⁺ gene carried by the λdnrd⁺ prophage enables the cell to form a colony. If the prophage is lost from a cell, which does occur on occasion, that cell will be unable to form a colony. Growth on this agar is not, however, sufficient to ensure that the prophage is present because a cell lacking the prophage might revert to the Nrd⁺ phenotype by acquiring a second mutation. Thus, it is worthwhile to test for the presence of a prophage. This can be done by selecting colonies that form on an RA/Hyu plate (in this laboratory period) and replating them on two plates, one of which is incubated at 20–30°C and one at 42°C (next laboratory period). Both prophages contain a mutation that causes induction of the prophage at 42°C; thus, JF413 should grow at 20–30°C, but not at 42°C. This will be considered to be an adequate criterion for the presence of JF413, although it should be noted that the possibility of both having lost the λdnrd^+ prophage and reverting to the Nrd⁺ phenotype has not been ruled out. JF335 also contains a prophage with the temperature-sensitive mutation in the repressor gene, so the same test (20–30°C growth vs 42°C nongrowth) will be used in testing this strain.

II. MATERIALS PER TEAM

1. Slant culture of *E. coli* JF335 (4-1)*

2. Slant culture of *E. coli* JF413 (4-2)

3. Slant culture of *E. coli* JF417 (4-3)

4. Slant culture of *E. coli* JF427 (4-4)

5. Three RA plates (3-2)

6. One RA/Hyu plate (3-2/Hyu containing 0.75 mg Hyu/ml)

III. PROCEDURES

1. Streak cultures of JF417 and JF427 to obtain single colonies on an RA plate (as shown in Figure 2-1) and incubate the plate overnight at 37°C (See Note 2).

2. Streak a culture of JF413 to obtain single colonies on an RA/Hyu plate and incubate overnight at 30°C or at room temperature for two days.

3. Streak a culture of JF335 to obtain single colonies on an RA plate and incubate overnight at 30°C or at room temperature for two days.

4. Prepare agar plates for use in Period 2.

---

*See Appendix A for the composition of the solutions and reagents used in each period. Each reagent or solution has a number, in parentheses, that corresponds to its placement in Appendix A.

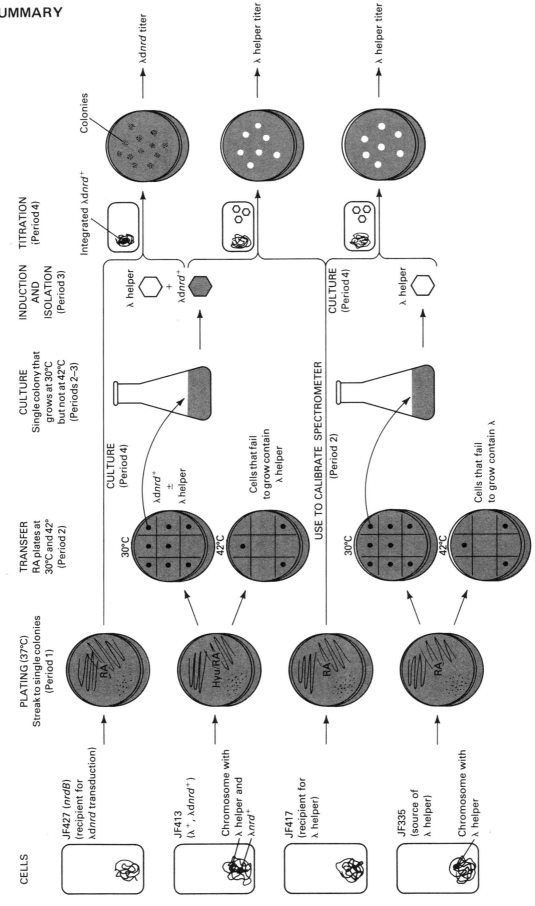

Figure P1-1 Relationship of the *E. coli* strains used to characterize λ helper and λd*nrd*⁺ phages.

54

V. NOTES

1. Be sure to reflame your streaking loop *and cool it* each time you change the directions of the streaking (see Figure 2-1). If you fail to reheat the loop, you will not get single colonies.

2. Place the plates upside down in the incubator—with the agar layer at the top and the lid at the bottom—so water droplets formed by condensation do not fall on the agar surface. Such droplets would cause smearing of the colonies, because daughter cells formed from a single bacterium can drift through the liquid and ultimately be deposited at a distance from the original parent cell.

3. Cloning of cells and molecular cloning of recombinant DNA molecules are different procedures using different materials. However, the processes are related, because in both cases a particular population of identical cells or DNA molecules is derived from a single progenitor cell or DNA molecule.

VI. INTERIM BETWEEN PERIODS 1 AND 2

The instructor will inoculate one 50-ml culture with JF417 cells obtained from one of the single colonies purified by the streaking in Period 1. This culture will be incubated in a 250-ml flask at 37°C overnight with shaking.

Growth of an *E. coli* Culture and Calibration of a Spectrophotometer

I. INTRODUCTION

In this period, you will learn how to grow *E. coli* in a liquid medium. Sterile growth medium will be inoculated with a small sample obtained from a nongrowing (stationary phase) culture. You will follow the growth of the bacteria as the culture passes from the lag phase of growth through the exponential phase and finally into the stationary phase. The quantity of bacteria present in the growth medium will be determined by measuring the absorbance at 600 nm (A_{600} or simply A) of the culture, using a spectrophotometer. These measurements will be made every 10 minutes and the value of A will be plotted as the y coordinate on semilog paper; time will be the x coordinate. When the number of bacteria is increasing exponentially, five samples will be taken at 15-minute intervals; each sample will be diluted to achieve a concentration of about 1000 cells/ml, and 0.1-ml samples will be plated to determine the number of viable cells present in each sample. Figure P2-1 shows a typical absorbance curve and the point at which you should begin plating. This procedure appears simple. However, a great deal of care is required if good plating results are to be achieved. All measurements must be accurate and the platings must be smooth. The purpose of this laboratory exercise is to stress the need for careful, thoughtful measurements, as well as to produce a calibration curve.

Plating is accomplished by spreading the 0.1-ml sample uniformly on the agar surface using a glass spreading rod (spreader). After the liquid is

Figure P2-1 Typical absorbance curve for a diluted culture of bacteria. After recording the absorbance values for the culture as it enters exponential growth, the closed points, begin plating cells when the culture is at the stage shown by the open point. Do five platings and continue to monitor the culture by taking absorbance readings until the absorbance values level off.

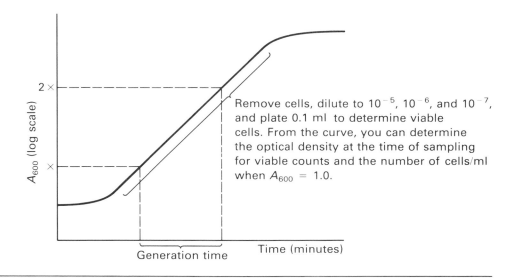

Remove cells, dilute to 10^{-5}, 10^{-6}, and 10^{-7}, and plate 0.1 ml to determine viable cells. From the curve, you can determine the optical density at the time of sampling for viable counts and the number of cells/ml when $A_{600} = 1.0$.

absorbed by the agar, each bacterium will remain in a single position and will multiply to form a visible colony. Thus, the number of colonies equals the number of bacteria in the sample; multiplying by ten (because only 0.1 ml is spread), and by the dilution factor, yields the number of bacteria per milliliter of original sample. For example, if 0.1 ml of the original sample is diluted into 10 ml of buffer (100-fold dilution), and 0.1 ml of this dilution is transferred to a second 10 ml of buffer (second 100-fold dilution), the total dilution will be $100 \times 100 = 10^4$; if plating 0.1 ml of the second dilution yields 64 colonies, the cell concentration of the original sample would be $10^4 \times 10 \times 64 = 6.4 \times 10^6$ cells/ml.

A few comments are in order about the plating procedure. Since you are interested in obtaining colonies only from bacteria in the original sample, the spreader must be sterilized. This is accomplished by dipping the horizontal portion in ethanol, removing the excess ethanol by shaking, and then burning off the remainder with a Bunsen burner. The spreader should not be left in the Bunsen flame because it can become hot enough to kill the bacteria in the sample to be spread; heating by simply letting the alcohol burn off is adequate. After the flame burns out, cool the spreader by touching it to a region on the plate devoid of cells; then, spread. Furthermore, it is a good idea to continue spreading the 0.1-ml sample until all liquid is absorbed. If plates having puddles of liquid are put in an incubator, some daughter cells produced in the first cell division might drift around and produce two or more colonies per initial cell deposited. In addition, plates should be put in the incubator upside down to avoid droplets that condense on the lid from falling on the agar surface and producing puddles. Also, liquid is often extruded from the uppermost surface of agar of freshly made plates and, if the plates are not inverted, the liquid will be on the same surface as the cells, and cause drifting.

From each sample taken from the exponentially growing culture, you will have a value of the absorbance and of the cell concentration. These pairs of values are to be plotted on semilog paper, to obtain a calibration chart for the spectrophotometer. This chart will be used many times in this course to determine the cell density simply from an absorbance reading.

A critical look at your plates after incubation will give you an indication of the quality of your sterile technique. Most microorganisms produce colonies having a characteristic shape. For example, *E. coli* colonies are flattened, creamy-white circles about 2 mm across with slightly fuzzy edges. Colonies of contaminating bacteria may be colored, rounded, very large or small, glossy, and so on, and colonies of fungi are usually exceedingly fuzzy. If your colonies are uniform in appearance, you have probably not introduced any contaminants. It is worth noting the odor of your plates. *E. coli* colonies have a mild odor, whereas some of the bacilli have very unpleasant and strong smells. For example, *Lactobacillus* smell of sour milk and *Pseudomonas* are quite aromatic and incenselike.

II. MATERIALS PER TEAM

1. A stationary culture of *E. coli* JF417 grown in RB medium (3-1) from the preceding interim period—one culture for the entire class

2. 40 ml of RB medium (3-1) in a 250-ml flask

3. Nineteen RA plates (3-2)

4. Twenty sterile tubes for dilutions (25-ml tubes work well)

5. 100 ml of RB medium (3-1) for dilutions

6. A large supply of sterile 1-ml, 5-ml, and 10-ml pipettes

7. A propipette or other pipette aid (see Appendix B)

8. A sheet of three-cycle semilog paper

III. PROCEDURES

1. Remove 1 ml of the stationary culture of JF417 with a sterile pipette and transfer it to 40 ml of prewarmed (37°C) RB medium contained in a 250-ml flask. Place the flask on a 37°C shaker.

2. For the next 180 minutes, at approximately 10-minute intervals, remove 1-ml samples and read the absorbance at 600 nm (A_{600}). To do so, adjust a blank sample to read zero and read the absorbance of this medium. Unless the value is zero, it must be subtracted from all absorbance readings of the culture. It is often convenient to use water as a blank, in which case the absorbance value of the medium will be greater than zero. If the A_{600} value is above 0.7, it will not reflect the true cell density, so the sample should be diluted in sterile RB medium and then read again. Remember to take into account this dilution when recording the absorbance.

3. Plot the values of A_{600} (y axis) at various times (x axis) as the data are obtained. This will enable you to choose an appropriate dilution for each subsequent reading.

4. When the culture has reached the exponential phase of growth, take five samples at 15-minute intervals. Read the absorbance of each sample and also dilute 0.05 ml into 4.95 ml of RB medium as the first 100-fold dilution in preparation for plating. Use a 200-μl Pipetteman for samples of 0.1 ml (100 μl) or less.

5. Do a second 100-fold dilution (total dilution, 10^4) and two consecutive 10-fold dilutions (10^5 and 10^6) by diluting 0.1 ml into 0.9 ml. See page 115 for an example of a dilution series.

6. Plate 0.1 ml from the 10^4, 10^5, and 10^6 dilutions. Be sure to write the dilution factors on the bottoms of the plates so, if the lids come off, your plates will not get mixed up. Put all plates in the 37°C incubator.

7. Using the straight-line portion of your curve (which indicates exponential growth), calculate the time interval required for the value of A_{600} to double; this is the generation time of the culture.

8. After 16–24 hours of incubation, count the colonies on the plates, using the dilutions having 50–500 colonies. If you have no dilutions with fewer than 500 colonies, count one-half or one-fourth of the colonies on a plate. (This is not particularly accurate but will provide an estimate.) Calculate the number of viable cells per milliliter of culture (the cell density) and then the ratio of the cell density to A_{600}. Calculate an average value of the cell density corresponding to $A_{600} = 1.0$.

The following steps use the streak plates prepared in Period 1.

9. Mark the bottom of two RA plates to divide the agar surface into eight sections (number the sections) and then transfer eight colonies from the JF413 plate to corresponding positions on the two plates. Incubate one plate at 20–30°C and one at 42°C.

10. Repeat Step 9 with JF335.

11. Prepare agar plates for Period 3.

IV. SUMMARY

The experiment is summarized in Figure P2-1. Plating and corresponding readings of A_{600} are made with samples taken from the portion of the growth curve indicated by the bracket.

V. NOTES AND QUESTIONS

1. Calibrate the spectrophotometer that you used for your experiments. What is the number of cells per milliliter if $A_{600} = 1.0$?

2. What is the generation time of your culture, determined from the spectrophotometer readings?

3. What is the generation time of your culture, determined from the platings you performed? If there is a difference between the values obtained by plating and by spectrophotometry, explain the difference.

4. At what cell density (A_{600} and cells/ml) does your culture leave the log phase (this is the last point on the straight-line portion of your curve)? You will need to know this value in future experiments.

5. What is the cell density (A_{600} and cells/ml) of your culture when it is in the stationary phase? How does this compare with the value for your overnight culture?

VI. INTERIM BETWEEN PERIODS 2 AND 3

1. The instructor will pick cells of JF413 from one streak that grew at 20–30°C, but not at 42°C, and then inoculate 40 ml of SB medium (3-3) contained in a 250-ml flask. This culture will be grown overnight at 30°C. Two hours before class, the instructor will dilute the culture 10-fold by adding the 40 ml to 360 ml of SB medium in a 1-l flask, and will put the culture on a shaker at 30°C. This single culture will be used by the entire class.

2. One culture of JF335 will be prepared in a similar manner except that 10 ml of an overnight culture will be diluted into 90 ml of SB medium in a 250-ml flask.

Note: The procedures for Periods 3 and 4 were designed for a laboratory class with six teams of two, and a small centrifuge with an eight-place rotor (Sorvall SS-34). Accordingly, eight samples of λ DNA can be prepared. We suggest that each team prepare λd*nrd* DNA and that two teams volunteer to prepare λ helper DNA for the entire class. For classes with fewer teams or additional centrifuge space the number of samples of λ helper in JF335 can be adjusted appropriately.

Growth of
Bacteriophage λ

I. INTRODUCTION

This period will require about five hours to complete, but most of the work will be concentrated at the beginning and the end of the period, so you should plan accordingly. The goal is to have a supply of both $\lambda d nrd^+$ DNA and λ helper DNA. Both phages are produced by JF413, but the procedure used to purify the phages will not separate the two. Therefore, the helper λ will be prepared independently from JF335.

The procedure for growing phage from a lysogen is to derepress the prophage and thereby initiate a lytic cycle of phage growth. This derepression is called *induction*. The prophages in these lysogens have a temperature-sensitive repressor, so to induce the cells you need only to raise the temperature of a growing culture to 42°C for 20 minutes. The culture could be kept at this temperature, but the yield of phage is improved by lowering the temperature to 37°C. (Induction is not reversed by lowering the temperature.)

All prophages contain the *S7* mutation, which prevents lysis of the bacteria during phage development. This mutation allows you to concentrate the cells by centrifugation prior to lysis. Thus, after three hours of phage development, the bacteria are concentrated by centrifugation and then lysed by the addition of chloroform.

The phage is suspended in a Tris buffer (ph 7.5) containing $MgSO_4$— the Mg^{2+} ion is needed to stabilize the phage. To aid in purification, the bacterial nucleic acids are removed from the lysate by adding a mixture of DNase and RNase; the phage DNA in the phage head is resistant to the DNase. Cellular debris is then removed by centrifugation, and the supernatant—which contains the phage particles—is stored at $4°C$; a few drops of chloroform are added to prevent the growth of bacteria.

It will be necessary to free the phage from all cellular DNA and most cellular protein. This will be done in Period 4.

Near the middle of this laboratory period, when you should not have too much to do, the instructor will give a brief lecture on the properties of plasmids; this will give you some background material for later laboratory exercises. You might also use some of your spare time to examine the plates you prepared in Period 2.

Two teams need to volunteer to prepare λ helper from JF335.

II. MATERIALS PER TEAM

1. Exponentially growing culture of JF413—one culture for the entire class will do; but, to be sure that the culture is at the proper dilution at the start of class, the instructor should have several different dilutions growing.

2. 50 ml of SB medium (3-3) supplemented with 0.5 ml of 1 M $MgSO_4$ (1-1)

3. One 250-ml sterile flask

4. One 50-ml chloroform-resistant (polypropylene) centrifuge tube

5. 5 ml of phage dilution (PD) buffer (pH 7.5) (1-2)

6. 1 ml of chloroform (2-1)

7. 10 μl of DNase at a concentration of 10 mg/ml (1-3)

8. 50 μl of RNase T1 at a concentration of 1000 units/ml (1-4)

9. One 15-ml Corex tube

10. A sheet of two-cycle semilog paper

Two groups will need a second set of all of the materials listed above plus an exponentially growing culture of JF335.

III. PROCEDURES

The procedures for preparing λd*nrd* phage from JF413 and λ helper from JF335 are identical.

1. Measure A_{600} of the exponentially growing culture of JF413. Determine the cell concentration from your calibration curve.

2. Add a sufficient volume of the JF413 culture to 50 ml of supplemented SB medium to yield a cell density of 5×10^7 cells/ml. Incubate the culture with shaking at 30° C.

3. Measure the absorbance every 10 minutes to be sure that the culture is growing exponentially; its generation time should be about 60 minutes.

4. When the cell density reaches 1×10^8 cells/ml, shift the culture to a 42°C water bath shaker for 20 minutes.

5. After 20 minutes at 42° C, transfer the flask to a 37° C water bath shaker and shake for three hours (see Note 2).

6. Chill the culture in ice water for five minutes.

7. Transfer the culture to a 50-ml centrifuge tube and centrifuge at $12,000 \times g$ (10,000 rpm in a Sorvall SS-34 rotor) for 10 minutes. (Appendix C contains a table listing values of centrifugal force as a function of rotor speed for other rotors.)

8. Pour off the supernatant and resuspend the pellet in 5 ml of PD buffer (pH 7.5).

9. Add 0.2 ml of chloroform, 10 μl of DNase, and 50 μl of RNase, and incubate for 20 minutes at 37° C.

10. Centrifuge at $12,000 \times g$ for 10 minutes.

11. Store the supernatant at 0–4° C until Period 4. Discard the pellet.

12. Prepare agar plates for Period 4 and store at 4° C.

IV. SUMMARY

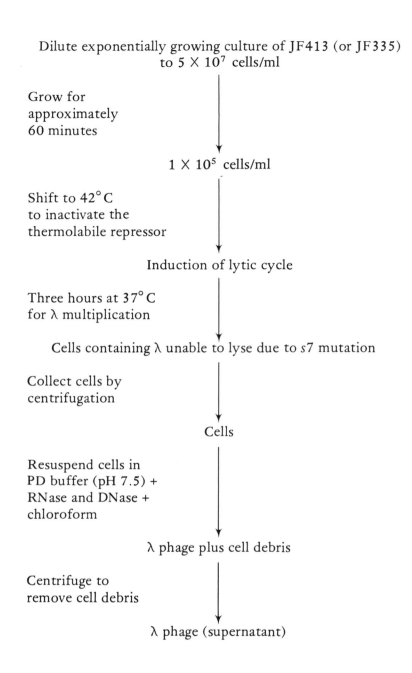

Dilute exponentially growing culture of JF413 (or JF335)
to 5×10^7 cells/ml

Grow for
approximately
60 minutes

1×10^5 cells/ml

Shift to 42°C
to inactivate the
thermolabile repressor

Induction of lytic cycle

Three hours at 37°C
for λ multiplication

Cells containing λ unable to lyse due to *s7* mutation

Collect cells by
centrifugation

Cells

Resuspend cells in
PD buffer (pH 7.5) +
RNase and DNase +
chloroform

λ phage plus cell debris

Centrifuge to
remove cell debris

λ phage (supernatant)

V. NOTES AND QUESTIONS

1. The actual number of cells per absorbance unit of JF335 and JF413 is larger than for JF417 but this has been taken into account in this experiment; the number of cells used is actually about twice as large as indicated by your calibration curve.

2. For maximal production of phage the cells should be in an active growth state, so that they will be synthesizing sufficient material to manufacture phage particles. Intracellular phage particles will accumulate for about three hours after the 42°C heat-shock. From your calculations from Period 2, calculate the increase in cell number that would have occurred if the culture had remained at 30°C. Will the culture still be in log phase growth at the end of the induction period?

VI. INTERIM BETWEEN PERIODS 3 AND 4

The instructor will start 50-ml cultures of JF417 and JF427 in MB medium (3-4) and incubate overnight at 37°C. These two cultures will be used by the entire class.

Titering of λ Helper Phage and Transduction of nrd^+ Genes with λdnrd^+

I. INTRODUCTION

In this period, you will concentrate the phage lysates prepared in Period 3 and determine the number of phage per milliliter of the original lysates.

Most phages are easily concentrated (and partially purified) by centrifugation because their sedimentation coefficients are higher than those of most cellular components. For example, phage λ has a sedimentation coefficient of 424S, whereas the values for the DNA and RNA fragments are no more than 20S and the values for proteins are generally less than 10S. Centrifugation of phage λ at 48,000 × g for three hours forms a pellet of phage λ that is free of most cellular components.

The number of phage per milliliter is determined by a plating technique. Phage are mixed with about 10^8 sensitive bacteria in a liquid agar (called *top agar* or *soft agar*). The liquid is poured on a solid agar surface and allowed to harden. The bacteria grow in the agar, converting it to a turbid layer called a *bacterial lawn*. Each viable phage reproduces in the agar while the lawn is forming and forms a clone of phage. As a result of this growth, a sufficient number of bacteria are lysed and a clear area in the lawn results; this clear area is called a *plaque*. Since a plaque is formed by one initial phage, plaque counting is a means of counting the number of phage in a liquid suspension; that is, aliquots from different dilutions are plated, as in the determination of cell density by colony formation. Remember that the λ helper phage possesses a mutation *S7* that prevents lysis; thus, *supF* bacteria, which suppress this mutation, are used to form the lawn.

Not all phage particles form visible plaques (but more than half do) for a variety of reasons, one of which is the very slow adsorption of phage λ to host bacteria. To improve the efficiency of plaque formation, the host bacteria are grown overnight in a maltose-containing medium. Maltose is a component of the λ adsorption site on the bacterial cell wall; addition of maltose to the growth medium increases the number of these sites and accelerates adsorption considerably. Preadsorption of phage to the cells also increases the efficiency of adsorption and increases the size of the plaques. Larger λ plaques can be obtained if the cells are not growing at their highest rates. Accordingly, maltose and glucose are omitted from the agar during the λ helper phage titrations. When pouring the liquid agar on the plates, it is necessary to tip the plate quickly so the liquid will cover the entire surface uniformly before hardening. If the agar solidifies during the movement of the liquid on the surface, the lawn will appear grainy and the plaques will be difficult to see. The formation of a grainy rough agar surface is a common problem for inexperienced microbiologists. As much as possible, try to avoid any delay in pouring the soft agar onto the plate.

The $\lambda \mathrm{d}nrd^{+}$ phage are defective and cannot form a plaque. They are counted by scoring the appearance of the nrd^{+} allele in a nrd^{-} host, that is, by transduction. A nrd^{-} bacteria cannot grow on agar containing hydroxyurea. However, a $\lambda \mathrm{d}nrd^{+}$ phage can inject its DNA into a $nrdB^{-}$ bacterium and its DNA can recombine with the host DNA to form a wild-type host, which can grow on agar containing hydroxyurea. This is not a particularly accurate way of counting the defective phage because the efficiency of transduction is fairly low—roughly 0.1% of the particles successfully form Nrd^{+} lysogens. However, for your purposes it will be sufficient to multiply the measured value of the transduction frequency by 1000 to determine the number of transducing particles per milliliter. The same experimental technique used for forming plaques is used to perform transduction except that the Nrd^{-} strain JF427 will be used and plated in soft agar containing hydroxyurea. Transduction will be detected by formation of small colonies within the agar. These colonies look much different from colonies on the surface of agar; they are about $\frac{1}{2}$ mm across and are lens-shaped.

II. MATERIALS PER TEAM

1. *E. coli* cells. Centrifuge 50 ml of the overnight cultures of JF427 (4-4) and JF417 (4-3) at 12,000 $\times$ g (10,000 rpm in the Sorvall SS-34 rotor) for 10 minutes and resuspend the two cell pellets in 25 ml of MB (3-4) containing 0.25 ml 1 M MgSO$_4$ (1-1). Only one team needs to prepare these two cultures for the entire class, for more than enough will be available for plating of all phage lysates.

2. λ phage preparations from JF413 and JF335 (Period 3)

3. One 50-ml polypropylene centrifuge tube. Two teams will need an additional 50-ml polypropylene centrifuge tube

4. 120 ml of PD buffer (pH 7.5) (1-2)

5. 24 ml of soft RA (3-2/s) without glucose or maltose, but containing 0.12 ml 1 M $MgSO_4$ (1-1)

6. Eight thick (30–40 ml) RA plates (3-2) without glucose or maltose, but containing 10 mM $MgSO_4$ (diluted from solution 1-1)

7. Nine ml soft RA/Hyu (3-2/s Hyu: use 0.75 mg hydroxyurea/ml)

8. Three RA/Hyu plates (3-2 Hyu: use 0.75 mg hydroxyurea/ml)

9. Twenty-six 25-ml dilution tubes

III. PROCEDURES

A. Harvesting of λd*nrd*$^+$ and λ Helper Phage from JF413

1. Centrifuge the phage preparations from the culture of JF413 (Period 3) for 10 minutes at 12,000 $\times g$ (10,000 rpm in the Sorvall SS-34 rotor) to remove residual cell debris.

2. Remove the supernatant, which contains the phage and which we call the *unconcentrated phage suspension*, with a sterile pipette, recording the volume. Be careful not to get any cell debris or chloroform in the pipette.

3. Add 0.1 ml of each unconcentrated phage suspension to separate sterile tubes for titration and transduction (later in this period). Store the samples in ice.

4. Put the remainder of each phage suspension in a 50-ml polypropylene tube and centrifuge at 48,000 $\times g$ for three hours (20,000 rpm in the Sorvall SS-34 rotor).

5. As soon as the centrifuge rotor stops (*Do not use the brake*), pour off the supernatant and add 0.4 ml of PD buffer (pH 7.5) to the tube. *The phage in the pellet rapidly resuspend so it is important that the supernatant be removed immediately* (see Note 1).

6. Allow the phage to resuspend until the next period by placing the tube at 4°C without shaking. (Vigorous resuspension can break the tails off, resulting in nonviable phage and often the release of the DNA.) After resuspension, you will have a concentrated phage stock. Calculate the expected phage concentration from the volume reduction and the titer of the unconcentrated phage suspension.

B. Harvesting of λ Helper Phage from JF335

1. Two groups need to repeat steps A-1 through A-6 using the culture of JF335 from Period 3.

C. Titration of Helper Phage in the Suspension Containing λd*nrd*$^+$

1. Add 0.1 ml of the unconcentrated phage suspension from the JF413 culture (step A-3 above) to 9.9 ml of PD buffer (pH 7.5) (100-fold dilution). Do one more 100-fold dilution and then make a series of 10-fold dilutions—10^4 to 10^{10}—in PD buffer. You should have eight dilution tubes at this point.

2. Add 0.1 ml each of the 10^7–10^{10}-fold dilutions to four 0.1-ml aliquots of a fresh overnight culture of JF417 prepared in (1) of Materials per Team. Label the tubes with the dilution factors. Save the 10^4–10^6-fold dilutions for Step E1.

3. Incubate the four tubes (10^7–10^{10} dilutions) with JF417 cells for 10 minutes at 37° C.

4. To each tube add 3 ml of melted soft agar (kept at 50–53°C) and mix by gentle swirling (two seconds). Do not swirl so hard that bubbles form because you will have difficulty distinguishing a bubble from a plaque.

5. *Rapidly* pour the contents of each tube on an RA agar plate and tilt the plate back and forth (as shown in Figure P4-1) so the soft agar covers the plate uniformly. Plates prewarmed to 37°C work best.

6. After the agar has hardened, incubate the five plates overnight at 37°C.

7. The next day, count the plaques and calculate the phage titers (plaques per milliliter of unconcentrated phage suspension).

Figure P4-1 Spreading soft agar on the hard agar plate.

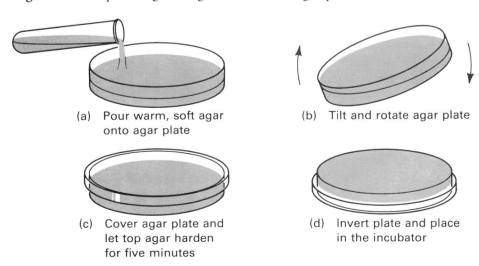

(a) Pour warm, soft agar onto agar plate

(b) Tilt and rotate agar plate

(c) Cover agar plate and let top agar harden for five minutes

(d) Invert plate and place in the incubator

D. Titration of Helper Phage from JF335

1. The groups that prepared the phage from the JF335 culture will add 0.1 ml of the unconcentrated phage suspension to 9.9 ml of PD buffer (pH 7.5) (a 100-fold dilution).

2. All of the teams in the class will use these two initial dilutions to make further dilutions, following steps C1–C7.

E. Transduction Assay of λd*nrd*⁺

1. Using the phage dilutions prepared above (step C-2), combine 0.1 ml of JF427 and 0.1 ml of the 10^2, 10^4, and 10^5-fold dilutions in three small labeled tubes.

2. Incubate for 30 minutes at $30°C$.

3. Add 3 ml of RA/Hyu soft agar (kept at $50–53°C$) to each 0.1 ml dilution, mix by gently swirling (two seconds), and pour the contents of each onto a separate RA/Hyu plate.

4. After the agar has hardened, place the three plates upside down in a $30°C$ incubator for 48 hours.

5. After colonies (*transductants*) have formed, count the colonies and calculate the number of transducing particles per milliliter of unconcentrated phage suspension.

IV. SUMMARY

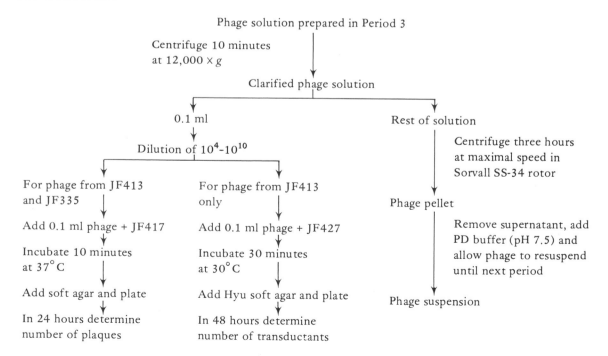

V. NOTES AND QUESTIONS

1. You will get a pellet of debris, which will not resuspend easily. The phage resuspend rapidly so you should be near the centrifuge when it stops. Remove the supernatant immediately.

2. What is the titer of the λ helper phage from JF413 and JF335 (determined from plaques formed on JF417)?

3. Assuming that you had about 5×10^8 JF413 and JF335 cells/ml at the end of the induction period in Period 3, how many λ helper phage did you obtain per cell?

4. The $s7$ mutation should enable you to obtain about 1000 phage particles per cell. How do your yields compare to this value?

5. How many micrograms of λ helper DNA do you have from each preparation? (The molecular weight of λ DNA is about 3.1×10^7.)

6. Assuming a transduction frequency of 10^{-3}, how many λdnrd phage particles do you have?

7. Assuming that you had about 5×10^8 JF413 cells/ml at the end of the induction period, how many λdnrd phage particles were produced by each cell?

8. How many micrograms of λdnrd do you have in your phage preparation? (The molecular weight of λdnrd is about 3×10^7.)

VI. INTERIM BETWEEN PERIODS 4 AND 5

1. The instructor will remove the plates with the λ plaques from the incubators after 24 hours and store them upside down at $4°$C.

2. The instructor will remove the plates containing the transductants from the incubator after 48 hours and store them upside down at $4°$C.

3. The instructor will begin an overnight culture of JF428 (4-5), which contains plasmid pBR325, the night before Period 5. The culture will be grown in RB medium (3-1) containing tetracycline and chloramphenicol.

Period 5

Plasmid Amplification and Preparation of λ DNA

I. INTRODUCTION

In this period, you will perform two procedures: (1) cells will be grown from which plasmid DNA will be isolated in another period, and (2) λ DNA will be isolated from phage λ.

The first procedure will use JF428, a strain of *E. coli* containing the plasmid pBR325. An overnight culture grown in the presence of tetracycline and chloramphenicol (to ensure that the plasmid has not been lost) will be diluted into fresh medium and the new culture will be allowed to grow to the late exponential phase—a cell density of about 5×10^8. The antibiotic spectinomycin will then be added to inhibit protein synthesis (see Note 1). Without protein synthesis *E. coli* cannot initiate a new round of DNA replication, although molecules in the act of replication can be completed. Replication of pBR325 continues under these conditions, because it is a relaxed plasmid and hence initiation does not require protein synthesis (or at least the tiny amount of protein made in the presence of spectinomycin is sufficient). Such continued replication of plasmids is called *plasmid amplification*. The antibiotic is added to cells in the late exponential phase in order to maximize the amount of plasmid DNA recovered per milliliter of culture. You must be careful that the culture has not entered the stationary phase because then little or no plasmid replication will occur. To avoid this potential problem, the cell density will be monitored by reading the absorbance of the culture several times per hour and determining the cell density from the calibration curve prepared earlier in Period 2. Following the addition of spectinomycin the culture will be incubated overnight. The next day, the cells will be harvested by the instructor.

Since the cells will be growing for several hours and absorbance readings can be taken quickly, you will use that time for isolating $\lambda dnrd^+$ DNA. This DNA will be used in Periods 7 and 8 as a source of the nrd^+ genes for cloning in the plasmid DNA.

II. MATERIALS PER TEAM

1. Overnight culture of JF428 (4-5); one culture for entire class.

2. Phage preparations from Period 4 (steps A-6 and B-6)

3. 70 ml of RB medium (3-1)

4. 0.5 ml of 100× spectinomycin (30 mg/ml) (1-5)

5. 125-ml Erlenmeyer flask

6. Several 5-ml disposable tubes for collecting 1-ml culture

7. A sheet of two-cycle semilog paper

8. 1.5 ml of STE-saturated phenol (1-7)

9. 1 ml of chloroform (2-1)

10. 50 μl of 2 M NaAc (pH 6.5) (1-8)

11. 1 ml of 95% ethanol (2-3)

12. Six 1.5-ml microfuge tubes

13. Two additional sets of materials 8-12 for groups preparing λ helper phage

III. PROCEDURES

A. Growth of JF428

1. Add 0.2 ml of the overnight culture to 1.8 ml of RB medium.

2. Measure the A_{600} of this dilution.

3. Determine the cell density of the overnight culture from the absorbance calibration chart. Remember the 10-fold dilution in Step 1.

4. Add an aliquot of the overnight culture to 55 ml of RB medium to achieve a cell concentration of about $1\text{-}2 \times 10^7$ cells/ml.

5. Remove 1 ml of the diluted culture prepared in Step 4 and measure the A_{600}. Repeat with a new 1-ml aliquot every 20–30 minutes. Discard the sample after determining the absorbance.

6. Record the A_{600} values and plot them as a function of growth time on semilog paper.

7. Estimate the cell-doubling time of the culture and the time at which the culture will reach a concentration of 0.8×10^8 cells/ml.

8. When the cell concentration is 0.8×10^8 cells/ml add spectinomycin (30 mg/ml) to yield a final concentration of 0.3 mg/ml.

9. Incubate overnight at $37°$C.

10. Tomorrow the instructor will stop the growth of the culture.

B. Isolation of λdnrd⁺ and λ Helper DNA

1. Transfer the phage solution (from Step A-6, Period 4) to a 1.5-ml microfuge tube and add an equal volume of STE-saturated phenol (see Note 2).

2. Rock the solution for five minutes at slow speed.

3. Centrifuge the solution in a microfuge for two minutes to separate the aqueous and phenol phases.

4. Remove the aqueous phase (upper phase) with a micropipette and transfer it to another microfuge tube.

5. Add 0.5 ml of a 1:1 mixture of STE-saturated phenol and chloroform and repeat steps 2–4.

6. Add 0.5 ml of chloroform and repeat Steps 2–4. The chloroform will remove the phenol dissolved in the aqueous phase.

7. Remove the aqueous (top) phase and place it in a microfuge tube.

8. Remove 20 μl and store at $-20°$C until Interim between Periods 6 and 7.

9. Add 50 μl of 2 M NaAc (pH 6.5) and 1 ml of 95% ethanol and vortex. Allow the DNA to precipitate at $-20°$C until Period 7.

10. Two groups need to prepare λ helper DNA (from step B-6, Period 4) by following steps B-1 through B-9 above.

IV. SUMMARY

A. Growth of Cells

Dilute an overnight culture of JF428 to 1–2 $\times$ 10^7 cells/ml.

Grow culture to a cell concentration of 0.8 $\times$ 10^8 cells/ml.

Add 0.5 ml of spectinomycin. Continue incubation.

B. Preparation of Phage DNA

Add phenol and mix for one minute.

Separate the aqueous and phenol phases by centrifugation.

Collect the aqueous phase.

Treat with phenol: chloroform.

Treat with chloroform.

Save a 20 μl sample.

Precipitate the DNA with NaAc (pH 6.5) and 95% ethanol.

V. NOTES AND QUESTIONS

1. It is common to use chloramphenicol to inhibit protein synthesis for plasmid amplification. In your experiments this antibiotic would be ineffective. Why?

2. Phenol is used to deproteinize nucleic acid solutions. The phenol denatures and precipitates the phage protein. Some protein is redissolved in the phenol phase but most protein remains at the interface between the phases. The nucleic acids are in the aqueous phase. For more information about the chemistry of procedures for purification of nucleic acids, see Marmur (1963) and Parish (1972).

VI. INTERIM BETWEEN PERIODS 5 AND 6

The instructor will transfer your cultures to 4°C 18–24 hours after addition of spectinomycin.

Preparation of Plasmid pBR325 DNA

I. INTRODUCTION

A strain of *E. coli* containing pBR325 was grown in Period 5. In this period, you will isolate plasmid DNA. The procedure of Ish-Horowitz and Burke (1981) with a few modifications will be used. This protocol provides, in a few hours fairly pure plasmid DNA free of most RNA and chromosomal DNA. The principle on which the technique is based is that lysis of bacteria at a high concentration by the addition of sodium dodecyl sulfate (SDS) produces a matrix of chromosomal DNA and cell debris and this aggregated mass can easily be separated from the cellular RNA and plasmid DNA by centrifugation. The supernatant that is obtained contains mainly nucleic acids of which 98–99% is RNA and 1–2% is DNA (most of which is plasmid DNA).

II. MATERIALS PER TEAM

1. Amplified culture of JF428 from Period 5

2. 0.4 ml of GTE buffer (pH 7.5) (1-9) containing lysozyme (4-18) at 1 mg/ml

3. 0.8 ml of lysis solution (1-10)

4. 0.6 ml of 5 M KAc buffer (pH 4.8) (1-11)

5. 10 ml of isopropanol (2-2)

6. 1.1 ml of TE buffer (pH 7.4) (1-12)

7. 1 μl of RNase T1 at a concentration of 1000 units/ml (1-4)

8. 0.5 ml of STE-saturated phenol (1-7)

9. 0.5 ml of chloroform (2-1)

10. 0.45 ml of 2 M NaAc (pH 6.5) (1-8)

11. 0.75 ml of 95% ethanol (2-3)

12. Three 1.5-ml minifuge tubes

13. One 50-ml centrifuge tube

14. One 15-ml Corex tube

15. Pasteur pipettes

III. PROCEDURES

1. Obtain your spectinomycin-treated JF428 culture and transfer it to a 50-ml centrifuge tube.

2. Centrifuge the cells at 8000 $\times$ g for five minutes at 4°C (8,000 rpm in a Sorvall SS-34 rotor).

3. Resuspend the cell pellet in 0.4 ml of GTE buffer (pH 7.5) containing 0.4 mg of lysozyme (add the lysozyme just before use). Transfer the cells with a Pasteur pipette to a 15-ml Corex tube. Maintain at room temperature for five minutes.

4. Add 0.8 ml of lysis solution. Mix gently (see Note 1) and incubate for five minutes on ice.

5. Add 0.6 ml of ice-cold 5 M KAc buffer (pH 4.8) (see Note 2) and mix gently; incubate for five minutes on ice.

6. Centrifuge the lysate for 10 minutes at 27,000 $\times$ g (15,000 rpm in a Sorvall SS-34 rotor) at 4°C, or at room temperature.

7. Remove the supernatant with a Pasteur pipette and place in a 15-ml Corex tube. Add 1 volume (about 2 ml) of isopropanol at room temperature.

8. Allow the nucleic acids to precipitate at room temperature for 10 minutes (see Note 3).

9. Centrifuge at 17,000 × g (12,000 rpm in a Sorvall SS-34 rotor) for 10 minutes at 4° C.

10. Pour off the isopropanol, blot the rim of the centrifuge tube with a paper tissue to remove the last drop of isopropanol. Dry the pellet in a vacuum for 10 minutes. Let air back into the vacuum chamber slowly to avoid blowing the precipitate out of the centrifuge tube. If a vacuum chamber is not available, gently direct a stream of air over the pellet. Be very careful not to blow your sample away.

11. Resuspend the pellet in 0.5 ml of TE buffer (pH 7.4) and transfer to a 1.5-ml minifuge tube.

12. Add 1 μl of RNase T1. Incubate the solution at 37° C for 30 minutes.

13. Add 0.5 ml of STE-saturated phenol and vortex for 30 seconds. This treatment removes the RNase and any other residual protein.

14. Centrifuge the DNA solution for two minutes in the microfuge to separate the aqueous (plasmid-containing) phase from the phenol (protein-containing) phase. There may be a layer of denatured proteins at the interface.

15. Remove the aqueous layer, which should be about 0.5 ml, with a 200 μl micropipette and place it in a second 1.5-ml microfuge tube. Add 0.5 ml of chloroform and vortex for 30 seconds. The chloroform will remove most of the phenol.

16. Centrifuge the solution for two minutes to separate the phases.

17. Collect the supernatant as in Step 15 and transfer it to a third 1.5-ml minifuge tube.

18. Add 50 μl 2 M NaAc (pH 6.5) and 0.5 ml of isopropanol. Vortex for two seconds and let the plasmid DNA precipitate for 10 minutes at −20° C.

19. Collect the precipitate by centrifugation for five minutes.

20. Remove the supernatant and resuspend the precipitated plasmid in 300 μl of TE buffer (pH 7.4).

21. Repeat Steps 18, 19, and 20.

22. Remove the supernatant and resuspend the plasmid pellet in 300 μl of TE buffer (pH 7.4). Remove 20 μl and store at $-20°$C. See Interim between Periods 6 and 7. Add 45 μl of 2 M NaAc (pH 6.5) and 750 μl (2.5 volumes) of ethanol. Allow the DNA to precipitate at $-20°$C until Period 7 (see Note 3).

Plasmid DNAs present at various stages of purification are shown in Figure P6-1 in Note 4.

IV. SUMMARY

Resuspend pellet of spectinomycin treated cells	0.4 ml GTE buffer (pH 7.5) with lysozyme
Add	0.8 ml 0.2 N NaOH, 1% (w/v) SDS
Incubate five minutes on ice	
Add	0.6 ml ice-cold 5 M KAc (pH 4.8)
Incubate five minutes on ice	
Centrifuge 27,000 $\times$ g, 10 minutes	
Recover supernatant— add	2.0 ml isopropanol
Incubate 10 minutes at room temperature	
Centrifuge precipate 17,000 $\times$ g, 10 minutes	
Remove isopropanol, dry pellet and resuspend	0.5 ml TE buffer (pH 7.4)
Add	1 μl RNase T1
Incubate 30 minutes, 37°C	
Remove protein, add	0.5 ml phenol

Vortex mix and centrifuge	
Recover aqueous phase	
Remove phenol, add	0.5 ml chloroform
Vortex mix and centrifuge	
Precipitate DNA, add	50 μl 2M NaAc (pH 6.5) 0.5 ml isopropanol
Incubate 10 minutes, $-20°$C	
Centrifuge precipitate	
Dry and resuspend pellet	300 μl TE buffer (pH 7.5)
Precipitate DNA, add	50 μl 2M NaAc (pH 6.5) 0.5 ml isopropanol
Incubate 10 minutes, $-20°$C	
Centrifuge precipitate	
Dry and resuspend pellet	300 μl TE buffer (pH 7.5)
Remove, store at $-20°$C	20 μl
Precipitate DNA, add	45 μl 2M NaAc (pH 6.5) 750 μl 95% ethanol
Store overnight, $-20°$C	

V. NOTES

1. The plasmid-containing cells are lysed *gently* to avoid shearing of the chromosomal DNA into many fragments. The plasmid DNA will be contaminated with the least amount of chromosomal DNA if the intact chromosomal DNA is pelleted with the cell debris (Step 6); small fragments of *E. coli* DNA inevitably will contaminate the plasmid DNA.

2. SDS is a powerful anionic detergent and will inhibit most enzymatic reactions as well as lysing cells and organelles. Consequently, it must be removed. Potassium dodecyl sulfate is insoluble. Thus, an excess of KAc is added.

3. Nucleic acids can be precipitated by adding either two volumes of 95% ethanol or 1 volume of isopropanol. Ethanol precipitates both DNA and RNA; in 50% isopropanol RNA is more soluble than DNA, which is quantitatively precipitated.

4. Photographs of two gels that illustrate some of the features of the system with which you are working are shown in Figures P6-1 and P6-2. The gel in Figure P6-1 shows the molecular species you will observe when electrophoresing plasmid DNA obtained at various stages of isolation. The electrophoretic procedure used is explained in Period 7. The DNA is moving from the top to the bottom of the photograph in four different *lanes*. The nucleic acids are made visible by the fluorescence of bound ethidium bromide. The lanes show the material obtained from four different steps (1) after cell lysis—Step 7; (2) after precipitation with isopropanol—Step 10; (3) after treatment with RNase—Step 20; (4) after precipitation with ethanol—Step 22. The second gel (Figure P6-2) contains various amounts

Figure P6-1 Stages of plasmid purification. The nucleic acids are identified by their binding of ethidium bromide.

Lane 1: Nucleic acids from lysed cells.
Lane 2: Nucleic acids after isopropanol precipitation.
Lane 3: Nucleic acids after RNase treatment.
Lane 4: Nucleic acids after two alcohol precipitations.

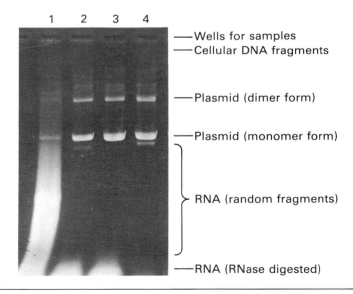

of plasmid DNA. The DNA was purified by centrifugation in a CsCl density gradient and then cleaved with *Eco*RI, so that it would form a single band. The seven lanes contain, in order, the following amounts of DNA: 5, 10, 20, 50, 100, 200, and 500 ng.

Figure P6-2 Agarose gel containing various quantities of plasmid. The DNA is stained with ethidium bromide. With this procedure, the limit of detection of a band of DNA is 2–5 ng.

Lane 1: 5 ng plasmid.   Lane 5: 100 ng plasmid.
Lane 2: 10 ng plasmid.   Lane 6: 200 ng plasmid.
Lane 3: 20 ng plasmid.   Lane 7: 500 ng plasmid.
Lane 4: 50 ng plasmid.

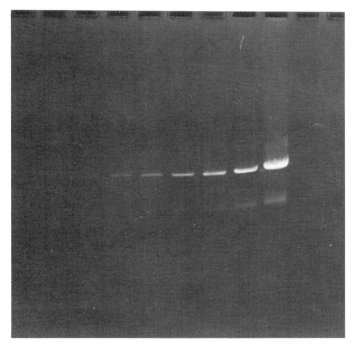

IV. INTERIM BETWEEN PERIODS 6 AND 7

The instructor will electrophorese the λ and pBR325 DNAs, from Periods 5 (step B-8) and 6 (step 22), on a 0.8% agarose gel in TAE buffer (pH 7.2); see Part F of Period 7 for a description of the gel. This gel will allow you to estimate the concentrations of DNA in your samples.

Period 7

Subcloning of Phage Genes into Plasmid DNA

I. INTRODUCTION

During this period, you will prepare recombinant DNA molecules by removing the nrd^+ genes from $\lambda d nrd^+$ DNA and inserting them into the plasmid DNA isolated during the last period. The procedure will be to digest the $\lambda d nrd^+$ DNA with the restriction endonuclease EcoRI and to join each of the five fragments to the single EcoRI site of plasmid pBR325, located in the cap gene. (In cloning jargon cleaving DNA with a restriction endonuclease is called **restricting**.) Since the nrd^+ genes are taken from a phage DNA molecule in which they had previously been cloned, the genes are being **subcloned**. The strategy for subcloning and the selection procedure used to identify pBRnrd recombinants is illustrated in Figure 4-5 (page 31). In future laboratory periods, you will examine various physical and chemical properties of the recombinant DNA molecules cloned in this period. Figure P7-1 shows an outline of the procedure to be followed.

A discussion of the handling of restriction enzymes is given in Appendix F.

Figure P7-1 Flowchart of procedures used to sublcone the *nrd* genes into pBR325.

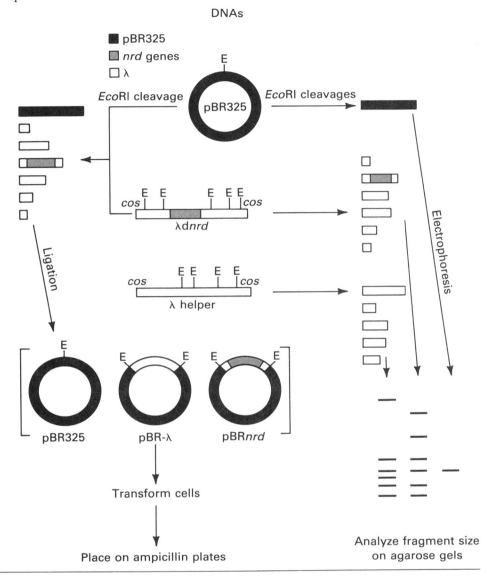

II. MATERIALS PER TEAM

      1. pBR325 DNA from Period 6

      2. λd*nrd*⁺ DNA from Period 5

      3. λ helper DNA from Period 5

      4. 1 ml of TE buffer (pH 7.4) (1-12)

5. 50 μl of 10× EcoRI buffer (pH 7.2) (1-13)

6. 40 units of restriction endonuclease EcoRI (4-12)

7. 50 ml of 20× TAE buffer (pH 7.2) (1-14)

8. One 50-ml flask

9. 0.3 g of agarose (2-4)

10. 15 μl of sample solution (1-15)

11. Ethidium bromide solution for staining (1-16); laboratory solution

12. 150 μl of isopropanol (2-2)

13. 6 μl of 10 mM ATP (1-17) freshly made and stored in an ice bath

14. 6 μl of 10× ligation buffer (pH 7.2) (1-18)

15. 15 μl of 2 M NaAc (pH 6.5) (1-18)

16. 2 units of T4 ligase (4-13)

17. Twelve 1.5-ml and 0.5-ml minifuge tubes

18. Several Pasteur pipettes

19. Micropipette tips

20. Whatman #3 paper

21. A sheet of two-cycle semilog paper

III. PROCEDURES

A. Determination of the Concentration of pBR325, λdnrd⁺, and λ Helper DNA.

1. Centrifuge the plasmid DNA prepared in Period 6 by centrifuging for five minutes in the minifuge. Aspirate the alcohol using the suction apparatus described in Note 1, dry the rim of the tube with absorbent paper, and dry the pellet in a vacuum for 15 minutes (see Note 2). Wait until you have prepared the λ DNA in Step 3 and dry the samples together.

2. Centrifuge the λ DNA samples in the microfuge for five minutes at 4° C.

3. Aspirate the alcohol, dry the rim of each tube and dry the pellets in a vacuum for 15 minutes (see Step 1).

4. Add 100 μl of TE buffer (pH 7.4) to each tube and resuspend the pellets.

5. Dilute 5 μl of each resuspended DNA sample into 1 ml of H_2O and determine both A_{260} and A_{280}.

6. Calculate the concentration of nucleic acids in each solution, using the relation: A_{260} = 1.0 corresponds to 50 μg DNA/ml (see Note 3). From the gel analysis of your λ and pBR325 DNAs, use Figure P6-2 to determine the concentrations of DNA in your preparations. Use these values below.

B. Restriction Digestion of pBR325 with *Eco*RI

1. Add H_2O to make 100 μl.

2. Add 10 μl of 10X *Eco*RI buffer (pH 7.2).

3. Add 1 μl (10 units) of *Eco*RI enzyme.

4. Add 5 μg of pBR325 DNA to a 0.5-ml minifuge tube. At this point the reaction mixture has the following composition:

___ μl pBR325 DNA (5 μg)

10 μl 10X *Eco*RI Buffer (pH 7.2)

1 μl *Eco*RI (10 units/μl)

___ μl water
_____
100 μl

5. Incubate at 37°C for 30 minutes.

6. Heat each reaction mixture to 75°C for five minutes to inactivate the *Eco*RI nuclease.

C. Restriction Digestion of λd*nrd*$^+$ with *Eco*RI Nuclease

Repeat Steps B1–6, replacing pBR325 DNA with λd*nrd*$^+$ DNA.

D. Restriction Digestion of λ Helper DNA with *Eco*RI Nuclease.

Repeat Steps B1–6, replacing pBR325 with λ helper DNA.

E. Restriction Digestion of a Mixture of pBR325 and λd*nrd*⁺ DNA with *Eco*RI Nuclease

Repeat Steps B1–6 using 2 μg of pBR325 DNA and 2 μg of λd*nrd*⁺ DNA in the same tube. The product of this reaction will be called {(pBR325 + λd*nrd*⁺) × *Eco*RI}.

The samples prepared in Sections B through E will each be used in the following steps. However, all of the sample will not be used and the remainder should be saved for future experiments.

F. Analysis of Restriction Digests of pBR325, λd*nrd*⁺, and λ Helper DNA

(See Appendix D for a detailed description of the gel electrophoresis apparatus.)

1. Prepare a 0.8% agarose gel by adding 0.2 g agarose to a 50-ml Erlenmeyer flask containing 25 ml H₂O, and boil until the agarose is completely dissolved. (Swirl the solution and look for undissolved chunks of agarose.) Then, add 1.25 ml of 20× TAE electrophoresis buffer (pH 7.2). Because some of the liquid is lost during boiling, you should still have about 25 ml.

2. Prepare a minigel by adding the hot (about 60°C) agarose to an 8.3 × 10.2 cm glass plate (Figure P7-2). Surface tension should hold the agarose on the glass unless the glass plate has nicks at the edge. Place a well-former in position and let the gel harden for 20 minutes; then, remove the well-former.

Figure P7-2 Preparation of an agarose minigel.

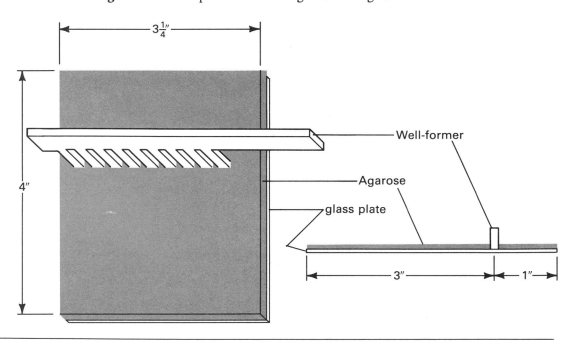

---

3. Remove a 5-μl sample (250 ng) from each reaction tube (procedures B, C, D, and E). Dilute each sample into a solution consisting of 8 μl of H_2O and 2 μl of sample solution contained in a 0.5-ml minifuge tube. Freeze the remainder of each reaction mixture, except that from E (pBR325 + λdnrd). Also, prepare two similar samples containing 250 ng of uncleaved plasmid DNA and uncleaved λ DNA.

4. Fill the reservoirs to capacity with 1× TAE electrophoresis buffer [diluted from 20× TAE buffer (pH 7.2)]. Add Whatman #3 filter paper wicks to complete the circuit between the electrodes in the reservoirs (Figure P7-3). Be sure to fill the reservoirs as full as possible. Most of the electrical resistance is across the filter paper. Consequently, if the paper bridge is too long, the DNA molecules in the gel will experience a lower effective voltage and will migrate much more slowly than if the paper connection between the gel and the buffer reservoir is short.

5. Place each DNA sample prepared in Step 3 into a single well in the gel.

Figure P7-3 Setup of the electrophoresis equipment.

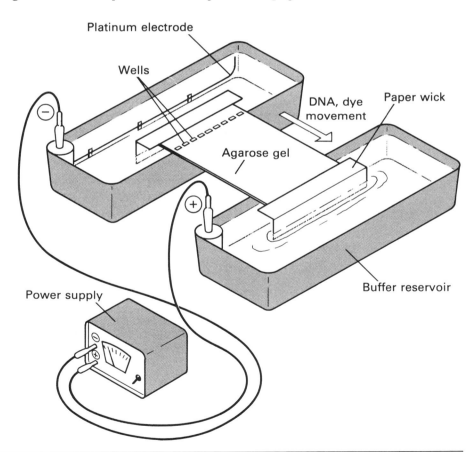

6. Electrophorese at 100–150 V/gel for 90–120 minutes, during which time the bromophenol blue marker should migrate two-thirds to three-quarters the length of the gel. Be sure to check soon after applying the voltage that the dye moves into the gel in the direction that the DNA fragments should move. If it does not, you have accidently reversed the polarity of the electric field. After the dye has moved into the gel, fill the wells with buffer, spread a thin layer of buffer over the entire gel and cover the gel plus the paper wicks with Saran wrap to keep the gel from drying. Also, keep an eye on the gel during electrophoresis to ensure that the wells do not dry up; if they do, the DNA in the lane will streak. If the temperature of the gel exceeds room temperature, lower the voltage.

7. When the dye has moved the appropriate distance, turn off the voltage and remove the glass plate. Stain the gel for 5–10 minutes by placing it in an ethidium bromide staining solution. This solution will be used by everyone and should remain in a covered dish. The electrophoresis buffer can be reused for each gel in the course if the contents of the reservoirs are mixed together to restore the pH.

8. Expose the stained gel to ultraviolet light and photograph the fluorescence of the stained gel (see Note 4).

9. Analyze the cleavage products found in the gel noting the completeness of cleavage, the λ helper fragment that is altered in λdnrd^+, and the multiple bands in the lane containing uncleaved pBR325 DNA (see Note 5).

G. Precipitation of 100 μl of *Eco*RI-cleaved (pBR325 + λdnrd^+)

1. Add 10 μl of 2 M NaAc (pH 6.5) to the reaction mixture of Step E-1.

2. Add 100 μl of isopropanol at −20°C.

3. Precipitate the DNA at −20°C for 20 minutes.

4. Centrifuge the DNA in a minifuge for 10 minutes. Place the tube in the minifuge with the hinges out; the pellet should be at the bottom of the tube on the hinge side, as shown in Figure P7-4.

5. Remove the isopropanol by aspiration, using a drawn-out Pasteur pipette.

6. Dry the pellet in a vacuum for 10 minutes.

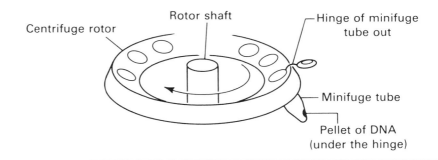

Figure P7-4 Pelleting of DNA precipitates in a minifuge tube.

H. Ligation

1. Resuspend the DNA pellet in 20 μl of H_2O. Make up the following reaction mixture:

10 μl Digested DNA

36 μl 10X ATP (10 mM)

36 μl 10X ligation buffer (pH 7.2)

__ μl T4 ligase (2 units)

__ μl Water

----

360 μl

2. Incubate the reaction mixture at 12°C overnight.

Save all DNA samples cleaved by *Eco*RI for reference markers in future experiments and 10 μl of {(pBR325 + λd*nrd*⁺) X *Eco*RI}.

IV. SUMMARY

Cleavage and analysis of phage and plasmid DNAs.

Precipate DNAs with isopropanol

Resuspend in 200 μl TE buffer (pH 7.4)

Digest 5 μg each DNA with *Eco*RI:

__ μl DNA

10 μl 10X *Eco*RI buffer (pH 7.2)

1 μl *Eco*RI (10 units/μl)

__ μl H_2O

----

100 μl

Incubate 30 minutes at 37°C

Heat five minutes, 75°C

Electrophorese digestion products
on a gel

Ligation of {(pBR325 + λdnrd^+) X EcoRI}

To {(pBR325 + λdnrd^+) X EcoRI} add   10 μl 2M NaAc (pH 6.5)
                                                        100 μl isopropanol

Precipitate, 20 minutes, −20°C

Centrifuge 10 minutes

Resuspend DNA fragments in 20 μl H_2O

Set up ligation reaction

    10 μl DNA

     6 μl 1 mM ATP

     6 μl 10X ligation buffer (pH 7.2)

   __ μl T4 ligase (2 units)

   __ μl H_2O
       ―――――
    60 μl

Incubate overnight at 12°C

Figure P7-5   Removal of alcohol supernatants from minifuge tubes.

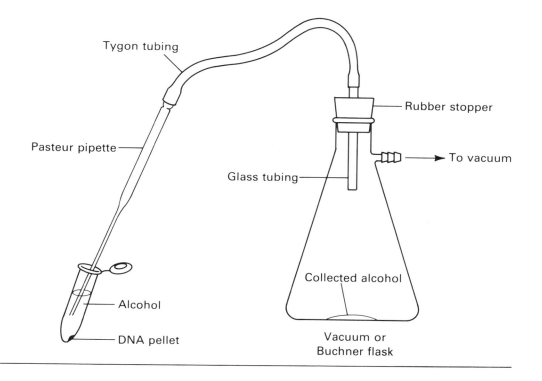

    PART TWO   LABORATORY EXERCISES

V. NOTES AND QUESTIONS

1. A vacuum aspirator for removing alcohol from minifuge tubes is shown in Figure P7-5. For removing the alcohol without disturbing the nucleic acid pellet, a $9\frac{3}{4}$-inch Pasteur pipette must be drawn, as shown in Figure P7-6. Heat the narrow part of the pipette evenly over a 2-cm length of the middle while rotating the pipette. When the area is soft, lift the pipette out of the flame and pull the ends apart; do not jerk the glass rapidly or pull too slowly. Break the pipette at the center of the drawn-out region and discard the leftover end.

Figure P7-6 Manufacture of drawn Pasteur pipette for the aspirator.

Figure P7-7 Air drying of nucleic acid precipitates.

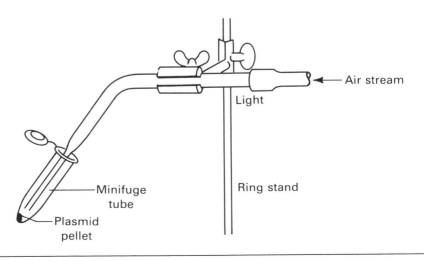

2. If a vacuum desiccator is not available, you can remove the alcohol remaining from the aspiration by directing a gentle flow of air over the pellet. Be very careful to avoid blowing the dried precipitate away. If too much alcohol remains, endonuclease cleavage and the ligation reaction will be inhibited. The setup is shown in Figure P7-7.

Figure P7-8 Ultraviolet light transilluminator and Polaroid camera setup. [From Maniatis et al. (1982).]

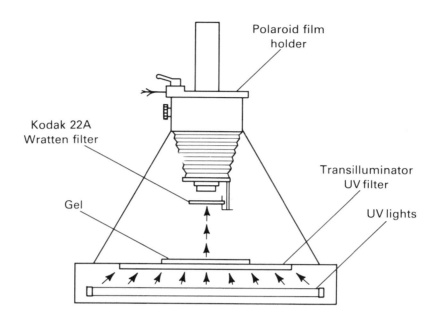

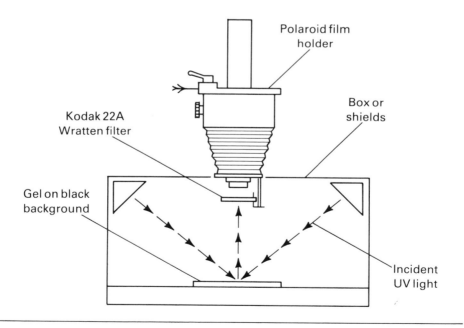

3. A solution of RNA having a value of $A_{260} = 1.0$ has a concentration of 45 μg/ml, nearly the same as for DNA. Pure DNA and RNA have a value of $A_{260}/A_{280} = 1.9 - 2.0$.

4. The bands of DNA are made visible by the fluorescence of ethidium bromide, which binds to the DNA by intercalation between the base pairs. If a Polaroid camera and an ultraviolet light transilluminator are not available (Figure P7-8), you can determine the positions of the gel bands by illuminating the gel with a hand-held ultraviolet light; however, the bands will not be as intense, because a transilluminator has a more intense light source.

5. The preparation of plasmid DNA should contain several species of DNA: supercoiled monomers, which should produce the most intensely fluorescing band; linear monomers, which form a faint band (if visible at all) slightly behind the supercoiled monomer; supercoiled dimers, which move more slowly than the monomers and may constitute as much as 30% of the DNA; nicked circular monomers, which result from random nicking of the supercoil during the alkaline and enzymatic digestion of the RNA that comigrate with the supercoiled dimers; and supercoiled multimers, which generally produce very faint bands higher up in the gel. RNA does not fluoresce as intensely as DNA, but you will be able to see RNA as a diffuse glow near the marker band of bromphenol blue. See Figure P6-1 for identification of the bands.

6. From the restriction enzyme maps given in Figure 3-2 of Chapter 3, determine the molecular weight values of the bands of the λ helper and $\lambda \mathrm{d}nrd^+$ DNA molecules cleaved with EcoRI. Determine whether there are any partial cleavage products—that is, products that contain two or more adjacent EcoRI fragments. Appendix D shows how DNA fragments migrate in agarose gels and how you can determine the sizes of the fragments.

7. From the standards in your manual (Figures P6-1 and P6-2 of Period 6) estimate (a) the approximate percentages of the various forms of plasmid DNA, (b) the amounts of DNA and RNA in your plasmid sample, and (c) the actual amounts of pBR325 cleaved with EcoRI that you added to the ligation reaction.

VI. INTERIM BETWEEN PERIODS 7 AND 8

1. The instructor will remove the ligation reactions, add 40 μl 2M NaAc (pH 6.5) and 1 ml ethanol, and store them at $-20°$C until Period 8.

2. The instructor will inoculate a culture of JF427 in 5 ml of RB medium (3-1) at $30°$C. The culture will be diluted prior to Period 8, so it will be actively growing at a density of about 5×10^7 cells/ml when the class begins. A volume of 50 ml is sufficient.

Period 8

Transformation of *E. coli* with Recombinant DNA

I. INTRODUCTION

In the first part of this period, you will examine the sizes of the products of the ligation reaction carried out in Period 7 in order to understand the relative rates of intra- and intermolecular joining (see Appendix E). In the second part of the period, you will transform strain JF427, which is nrd^- and hence cannot grow on plates containing hydroxyurea, with the recombinant DNA molecules prepared in Period 7. Hanahan (1983) discusses the effects of various components in the transformation procedure. We use a simple but very effective protocol in this experiment. The phenotype of the transformed cells will depend on the particular DNA molecule that enters the cell, as shown earlier in Figure 4-5 (page 31). Growth conditions after transformation can select either for or against a particular phenotype. Thus, since efficient growth of JF427 requires the expression of the *nrd* genes, a disproportionate percentage of viable transformed cells (as measured in your experiments) might be expected to contain a plasmid DNA molecule with the *nrd* genes *in a functional state*. We call these recombinant plasmids pBR*nrd*. However, smaller plasmids are more efficiently taken up by cells during the transformation process, so it is not possible to predict the number of transformants containing the *nrd* genes.

In the transformation experiment you will perform, cells will be transformed with recombinant plasmid DNA molecules. Normally, two important control experiments would also be carried out in parallel transformations, although you will not do this in your experiment. Uncleaved plasmid DNA would be used to verify that the cells are in fact able to be transformed, since if the experiment performed in this period were to yield no transformants, you could not distinguish the possibility of defective recombinant DNA molecules from cells unable to be transformed. Cleaved but nonligated DNA would be used in the second control to assay the number of clones derived from uncleaved plasmid molecules.

Restriction analysis is a commonly used method for determining whether recombinant plasmid molecules have been formed. However, even though transformation is not 100% effective, the transformation assay is several orders of magnitude more sensitive for detecting recombinant DNA molecules than restriction analysis.

II. MATERIALS PER TEAM

1. *Eco*RI-cleaved DNA preparations from Period 7

2. Ligated DNA sample from Period 7

3. 20 ml of an actively growing culture of JF427 at a concentration of about 5×10^7 cells/ml

4. 5 μl of 2 M NaAc (pH 6.5) (1-8)

5. 55 μl of isopropanol (2-2)

6. 60 μl of TE buffer (pH 7.4) (1-12)

7. 50 ml of 20× TAE buffer (pH 7.2) (1-14); use 1.25 ml if the reservoir buffer has been recycled from Period 7

8. Ethidium bromide staining solution (1-15)

9. 0.3 g of agarose (2-4)

10. 100 μl of sample solution (1-15)

11. 7 ml of 0.05 M $CaCl_2$ (1-19)

12. 6 ml of Z broth (3-5)

13. Nine RA agar plates (3-2)

14. Twelve RA/amp agar plates (3-2; containing ampicillin)

15. Ten RA/(amp + Hyu) agar plates (3-2; containing both ampicillin and hydroxyurea at 0.75 mg/ml)

16. Five 0.5-ml minifuge tubes

17. Three 15-ml Corex tubes

18. An exponentially growing culture of JF427

III. PROCEDURES

A. Analysis of Recombinant DNA

1. Centrifuge the ligated DNA samples, from the interim period, in the minifuge for 10 minutes at 5°C.

2. Remove the supernatant by aspiration and dry the pellets in a vacuum for 10 minutes.

3. Resuspend the pelleted DNA in 10 μl of TE buffer (pH 7.4); this should yield a concentration of about 200 ng of DNA/μl. Keep the resuspended DNA samples on ice.

4. Prepare a 0.8% agarose minigel in 20X TAE buffer (pH 7.2) (see Procedures, Period 7).

5. Run the following samples (see Period 7) on the gel:

(a) 200 ng of (λdnrd^+ X EcoRI) digestion products

(b) 200 ng of (λ helper X EcoRI) digestion products

(c) 200 ng of (pBR325 X EcoRI) digestion products

(d) 500 ng of {(pBR325 + λdnrd^+) X EcoRI} ligation products

(e) 500 ng of {(pBR325 + λdnrd^+ I X EcoRI} —unligated from Period 7, step H.

(f) Uncleaved λdnrd, λ helper, and pBR325 markers—use amounts that are appropriate, based on the ease of visualization of the bands you observed in the gel in Period 7

6. Electrophorese the samples at 100–150 V/gel for 90–120 minutes, as in Period 7. Recycle the electrophoresis buffer, as you did in Period 7.

7. Stain the gel with ethidium bromide and photograph.

8. From the number of bands and the mobilities of each, estimate the extent of the joining reaction. Remember that for a particular molecular weight, linear molecules migrate faster than circular molecules.

B. Transformation with pBR*nrd* Recombinant DNA

Use ice-cold solutions for Steps 1–9.

1. Take 12 ml of an exponentially growing culture of JF427 at a concentration of 1–2 × 10^8 cells/ml (measure the absorbance and use your calibration curve) and centrifuge the cells at 5000 × g (6500 rpm in a Sorvall SS-34 rotor) for five minutes at 4°C in a 15-ml Corex tube. Discard the supernatant.

2. Resuspend the cells in 6 ml of 0.05 M $CaCl_2$ by flicking the tube with your finger. Divide into 3 2-ml aliquots in 15-ml Corex tubes. Put cells into an *ice bath* for 20 minutes.

3. Centrifuge the cells at 3000 × g for five minutes at 5°C. Discard the supernatants and drain the inverted tubes on absorbent paper for 30 seconds.

4. Resuspend the cells very gently in 0.2 ml of 0.05 M $CaCl_2$.

5. Add 7.5 μl (about 1.5 μg) of ligation products from Step A3 to tube 1. Add 7.5 μl (about 1.5 μg) of unligated plasmid DNA (Period 7, step H) to tube 2. Add 1.5 μg pBR325 to tube 3.

6. Maintain on ice for 40 minutes without shaking.

7. Put the tubes into a 42.5° C water bath for 90 seconds.

8. Add 1.8 ml of Z broth to each tube and incubate at 37°C for 90 minutes. This incubation increases the efficiency of transformation by allowing the transformed cells to recover from the abnormal ionic environment and to begin synthesizing the plasmid-encoded β-lactamase needed to overcome the ampicillin present in the RA/amp agar.

9. Dilute the transformed cells 10^6, 10^5, and 10^4-fold and plate 0.1 ml onto RA plates (two plates per dilution). The colonies that will grow on these plates will indicate the total number of viable cells after transformation.

10. Dilute the cells 10- and 100-fold and plate 0.1 ml of these dilutions and 0.1 ml of undiluted cells onto RA/amp plates (two plates per dilution). The colonies that will grow on these plates will reflect the total number of cells transformed with pBR325 (with or without inserted DNA).

11. Dilute the cells 10-fold and plate 0.1 ml of this dilution and 0.1 ml of undiluted cells onto RA/(amp + Hyu) plates (two plates per dilu-

tion). The colonies that will grow on these plates develop from cells transformed with recombinant plasmids that contain the *nrd* genes.

12. Centrifuge the remaining cells at 3000 × g for five minutes at 4°C. Discard the supernatant and resuspend the cells in 0.2 ml of Z broth. Plate 0.1 ml of these cells on each of 2 RA/(amp + Hyu) plates. This procedure should ensure that several pBR*nrd* recombinant plasmids are obtained. At this point all of the transformed cells have been plated.

13. Put all plates upside down in a 37°C incubator.

IV. SUMMARY

A. Analysis of Ligation Products

Ligation products	50 μl
Add 2 M NaAc (pH 6.5)	5 μl
Add isopropanol	55 μl
Precipitate: 20 minutes, −20°C	
Pellet: 10 minutes, 5°C	
Resuspend in TE buffer (pH 7.4)	10 μl
Electrophorese on a gel	1 μl

B. Transformation of JF427

JF427, 1–2 × 10^8/ml	12 ml
Centrifuge cells at 3000 × g	
Resuspend in 0.05 M CaCl$_2$	6 ml
Incubate 20 minutes on ice	
Centrifuge cells at 3000 × g	
Resuspend in 0.05 M CaCl$_2$	200 μl
Add DNA samples	5 μl
Incubate 40 minutes on ice	
Heat-shock 30 seconds, 42.5°C	

Add Z broth   1.8 ml

Incubate 90 minutes at 37°C

Plate on RA plates, 10^{-6}, 10^{-5}, 10^{-4} final dilutions

  RA/amp plates, 10^{-3}, 10^{-2} final dilutions

  RA/(amp + Hyu) plates, 10^{-2}, 10^{-1} final dilutions plus
    concentrated cells

V. NOTES AND QUESTIONS

1. For the ligation reaction of Period 7 calculate the value of j/i (see Appendix E) for the pBR325 DNA cleaved with *Eco*RI. How does this value compare with the value suggested by the graph in Appendix E? Do you expect more circular molecules than plasmid concatemers?

2. What is the value of j/i for the λdnrd^+ DNA fragment produced by *Eco*RI digestion that contains the *nrd* genes? Do you expect more circular molecules than concatemers for this fragment?

3. Assume that the average size of all plasmid and λ DNA fragments is about 6000 base pairs and that the molecular weight of one base pair is 660. Use the graph in Appendix E to determine the average value of j/i for the reaction between pBR325 and λdnrd^+ DNA. What types of products do you expect to obtain?

4. How do the products formed in the actual reaction compare with the predictions made from the calculation you just made?

VI. INTERIM BETWEEN PERIODS 8 AND 9

1. The instructor will remove the agar plates containing transformed JF427 from the incubator after the colonies are large enough to count and will store the plates upside down at 4°C.

2. The instructor will prepare plates of JF428 and JF429 that have single colonies for use in Period 9.

Period 9

Biological Analysis of Recombinant Plasmids

I. INTRODUCTION

Joining of *Eco*RI fragments of λd*nrd*⁺ DNA and linearized pBR325 DNA molecules yields four possible intramolecular products (see Note 1), 10 bimolecular intermolecular products, and many more multimolecular products (see Notes 2 and 3). Appendix E contains a discussion of one method for determining whether inter- or intramolecular products are favored. In this period, in order to identify the variety of products of the joining reaction and to isolate pBR*nrd* recombinant DNA for further use, you will use the biological properties conferred by the plasmids on their host to select various recombinant types. Recombinant plasmids will be designated pBR*nrd*.

Overnight cultures of several Amp^r transformed cells obtained from the plates of Period 8 will also be started; these will be used during Period 10.

During this laboratory period there will be time to reexamine by gel electrophoresis any of your DNA samples that were inadequately resolved in Periods 7 and 8.

During the spare moments in this period, read about *expression plasmids*—those engineered for generating large amounts of mRNA for translation into a protein of value. [See Old and Primrose (1982), Maniatis et al. (1982), and Bernard and Helinski (1980).]

II. MATERIALS PER TEAM

1. RA/amp plates (3-2/amp) containing transformed JF427 colonies from Period 8

2. Two RA/cap plates (3-2/cap)

3. Two RA/Hyu plates (3-2/Hyu)—use 2.0 mg of Hyu/ml (see Note 4)

4. Sterile toothpicks (in a foil-covered beaker)

5. 75 ml of RB medium (3-1)

6. 75 μl of 20 mg/ml tetracycline (1-21)

7. One plate with single colonies of JF428 containing pBR325

8. One plate with single colonies of JF429 containing pBR*nrd* recombinant plasmid

9. Fifty 5-ml sterile test tubes with caps

III. PROCEDURES

A. Biological Analysis of Transformed Cells

1. Obtain four selective plates—2 RA/cap and 2 RA/Hyu—and rule grid lines on these plates as shown in Figure P9-1. These plates will be

Figure P9-1 Grid patterns.

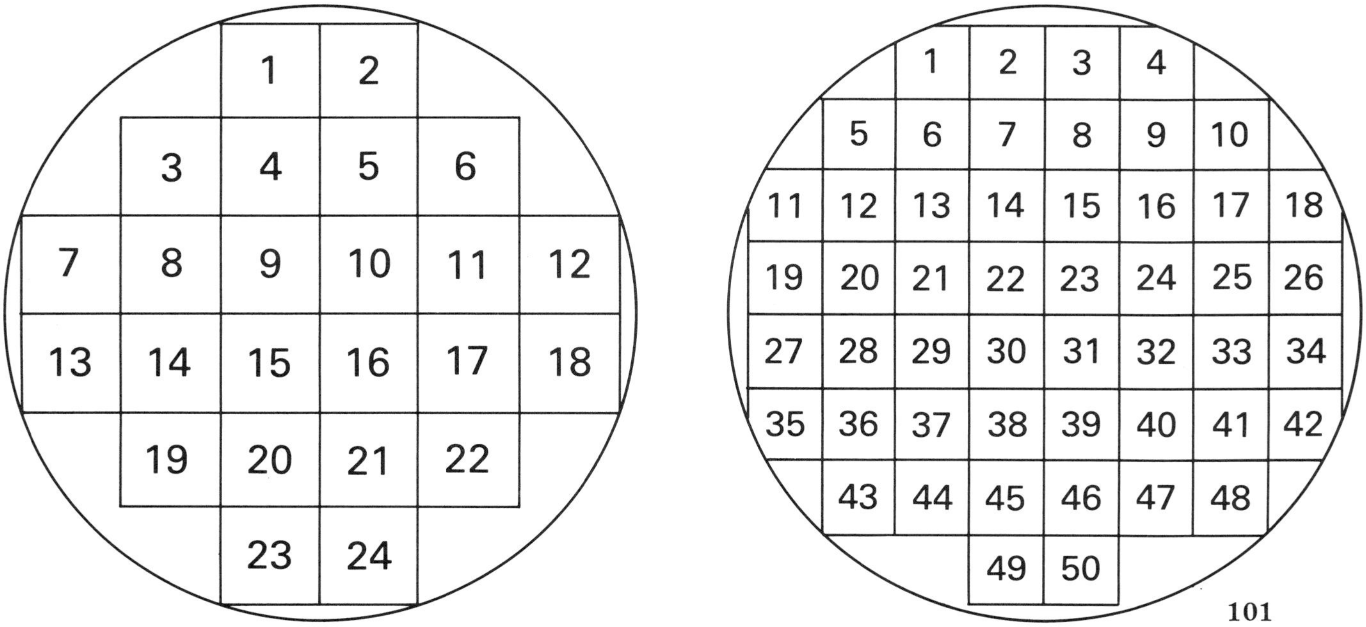

used for examining the phenotypes of particular clones of transformed cells.

2. Using a sterile toothpick, touch a colony on the RA/amp plate and transfer the cells first to the RA/cap plate and then to the RA/Hyu plate. Several colonies should be picked and each colony should be transferred to the same numbered positions on the two plates. Use a fresh toothpick for each colony. Transfer a few cells from one colony of JF428 containing pBR325, a few cells from one colony of JF429 with pBR*nrd* and a few cells from one colony of JF427, and plate these as controls. Pick five colonies from the RA/(amp + Hyu) plates to ensure that you will get cells with pBR*nrd* plasmids. Pick up as few cells from each colony as possible; too many cells will give a false positive result (see Note 5). You should plate at least 50 colonies, if possible.

3. Incubate the plates at 37°C overnight.

B. Physical Analysis of Transforming Plasmids

1. Using a sterile toothpick, transfer a few cells from each RA/amp colony, used in Steps 1 and 2, to a 5-ml culture tube containing 1.5 ml of RB medium supplemented with 20 μg of tetracycline/ml (see Note 6). Prepare 50 cultures if possible.

2. Incubate the tubes overnight at 37°C in a shaker or roller apparatus.

IV. SUMMARY

Test transformed JF427 for resistance to various drugs.

Prepare overnight cultures of transformed JF427.

Calculate the transformation frequencies.

V. NOTES AND QUESTIONS

1. The terminal fragments of λ DNA each have one *cos* end and one *Eco*RI end and hence cannot form inter- or intramolecular circles with fragments containing two *Eco*RI termini. However, one terminal fragment can form an intermolecular circle with another terminal fragment having the complementary end.

2. The number of combinations C (in which the order is not specified) of n restriction fragments (with the same termini) joined at a time is given by

$$C = \frac{(n + m - 1)!}{[m!(n - 1)!]}$$

Here C is the number of recombinant molecules of a given size resulting from each order of joining (m). For example, if three restriction fragments —a, b, and c—of discernible size are present, then three products (a, b, and c) can form from three unimolecular (intramolecular, $m = 1$) events; six products (aa, ab, ac, bb, bc, cc) from bimolecular (intermolecular, $m = 2$) events; and 10 products (aaa, aab, aac, abb, acc, bbb, bbc, bcc, ccc, abc) from trimolecular ($m = 3$) events. In a particular experiment, all of these products, as well as those containing 4, 5, 6, . . . , fragments, can form. Each combination has a particular size, which may or may not be resolved by electrophoresis.

3. The calculations in Note 2 do not take into account the polarity or specific sequence of a particular product. In the example in Note 2, the *abc* of the trimolecular reaction has four forms (Figure P9-2), having identical size and composition, but biological properties that might differ, depending on the orientations and relative order of the genes and regulatory elements. Thus, using specific biological properties to select particular recombinant plasmids from the total pool of products may not yield the proportions calculated by the method described in Note 2. In Period 10 this phenomenon will be pursued further.

Figure P9-2 A single trimolecular ligation product with four possible combinations of gene polarities.

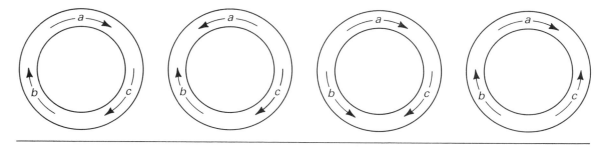

4. Several generations after transformation the number of plasmids per cell is higher. Initially, only one plasmid is present in each transformed cell. In successive generations the number increases to a constant number, which can be between 20 and 100; the actual number depends on the size and nature of the recombinant. Since the number of *nrd* genes will increase with the copy number of the plasmid, the concentration of Hyu can be increased in this and later experiments to ensure a stringent selection of *nrd*-containing recombinants.

5. Hydroxyurea inhibits ribonucleotide reductase and this in turn prevents DNA synthesis. *E. coli* continues to increase in mass (but not cell number) in the absence of DNA replication, and forms long nonviable filaments. If too many cells are placed on a Hyu plate, the increase in cell mass in these inhibited cells will appear to represent real growth and these cells would be classified (in error) as Hyu^r. This misleading result can be avoided by picking a number of cells that is small enough to ensure that the cell enlargement does not produce a visible spot of cells (apparent growth) on the agar.

Cells with pBR*nrd* would be expected to have many copies of the *nrd* gene and to be significantly more resistant than a wild-type strain. In fact, you could have isolated a pBR*nrd* clone without even using a *nrdB*$^-$ mutant strain. In this case, you would have allowed the transformed cells to grow and accumulate multiple copies of the recombinant plasmid before plating and selecting for cells having a Hyu^r phenotype. In your experiments, transformants containing only one copy of pBR*nrd* were plated.

6. Tetracycline is added to the medium to ensure that only plasmid-transformed cells grow. Tetracycline is used because all of the recombinant plasmids produced in Period 8 should have an intact *tet* gene.

7. Determine the number of viable cells in the transformation reactions from Period 8 from the number of colonies that grew on the nonselective RA plates.

8. Determine the number of transformed cells from the colonies that grew on the RA/amp plates. What are the transformation frequencies?

9. What is the number of transformed cells per microgram of recombinant DNA (which, for this calculation, you should assume to equal the number of micrograms of pBR325 DNA in the reaction mixture)?

10. How do you explain the presence of transformed cells from the JF427 to which unligated DNA fragments were added?

VI. INTERIM BETWEEN PERIODS 9 AND 10

The instructor will remove the RA/cap and RA/Hyu plates and store them at 4°C. The 1-ml cultures (Step 3 of Procedures) will also be removed and stored at 4°C.

Period 10

Size Analysis of Recombinant Plasmids

I. INTRODUCTION

In this period, you will complete the biological analysis of the recombinant plasmids initiated in Period 9. By combining all of the data obtained by various laboratory teams, a good statistical distribution of plasmid genotypes should result. The sizes of several recombinant plasmids will be analyzed by selecting samples of cells having various phenotypes, isolating the plasmid DNA molecules, and determining their sizes by agarose gel electrophoresis. A procedure in which several small samples of cells are utilized, called a *miniscreen*, will be used. In this procedure cells are lysed and, without purification, the lysate is examined for the presence of plasmid DNA. A large amount of RNA is present in the lysates but, after digestion with RNase, it will not interfere because plasmid DNA migrates much more slowly than cellular RNA (see Note 1). The miniscreen protocol is a shortened version of the procedure used in Period 6.

From the size of the DNA and the drug resistance genes it is possible to guess which *Eco*RI fragments have combined to form a particular recombinant plasmid.

Three phenotypes can be distinguished in the experiments of Period 9:

(a) $Amp^r Tet^r Cap^r Hyu^s$

(b) $Amp^r Tet^r Cap^s Hyu^s$

(c) $Amp^r Tet^r Cap^s Hyu^r$

All plasmids, recombinant or otherwise, should confer the Amp^{r}Tet^r phenotype. Cells containing recombinant plasmids with DNA inserted in the *cap* gene will also be Cap^s; those without inserted DNA are Cap^r. Plasmids with an *nrd*$^+$ gene inserted in the *cap* gene provide the Hyu^r phenotype but cells containing recombinant plasmids in which other fragments have been inserted will be Hyu^s.

The polarity of recombined restriction fragments can be important for gene expression (see Note 3 of Period 9). To determine if this is the case for the *nrd*$^+$ genes, the relative location of two restriction sites in several plasmids obtained from transformed JF427, having the Hyu^r phenotype, will be examined. These experiments will begin this period and be completed in the next period.

II. MATERIALS PER TEAM

1. Overnight cultures of transformed JF427 from Period 9

2. 1.6 ml of GTE buffer (pH 7.5) (1-9), containing lysozyme (4-18) at 1mg/ml

3. 3.2 ml of lysis solution (1-10)

4. 3.2 ml of 5 M KAc buffer (pH 4.8) (1-11)

5. 9.0 ml of isopropanol (2-2)

6. 1 ml of TE buffer (pH 7.4) (1-12)

7. 50 μl of 2 M NaAc (pH 6.5) (1-8)

8. 30 μl of sample solution (1-15)

9. 0.3 g of agarose (2-4)

10. 50 ml of 20× TAE buffer (pH 7.2) (1-14)—use 1.25 ml if recycled reservoir buffer is used

11. 40 μl of 10× Bam buffer (pH 8.0) (1-22)

12. 100 units of *Bam*HI endonuclease (4-14)

13. 10 μl of RNase T1 at a concentration of 1000 units/ml (1-4)

14. Thirty-two 1.5-ml minifuge tubes

15. Twenty-two 0.5-ml minifuge tubes

III. PROCEDURES

A. Biological Analysis of the Three Antibiotic Plates of Period 9

1. By observing the phenotypes of the colonies, determine the percentage of the plasmid-transformed cells in which a DNA fragment has been inserted at the $EcoRI$ site of plasmid pBR325.

2. Determine the percentage of plasmid-transformed cells and the percentage of recombinant plasmid-transformed cells that contain functioning nrd^+ genes. Record the colony numbers of the latter.

3. With your student colleagues prepare a chart showing the results for the whole class of the transformation experiment done in Period 9 (see Note 2).

B. Physical Analysis by Miniscreening

1. From the plating results in Steps A-1 and A-2, select 10 cultures containing plasmids with each of the following phenotypes:

(a) $Amp^r Tet^r Cap^r Hyu^s$ (three cultures)

(b) $Amp^r Tet^r Cap^s Hyu^s$ (three cultures)

(c) $Amp^r Tet^r Cap^s Hyu^r$ (four cultures)

2. Transfer the 10 selected cultures to 1.5-ml minifuge tubes and centrifuge for two minutes. Discard the medium.

3. Resuspend the cells in 0.1 ml of GTE buffer (pH 7.5) containing lysozyme at a concentration of 1 mg/ml. Incubate the cells for five minutes at room temperature.

4. Add 0.2 ml of lysis solution. Slowly invert the tubes three times to mix the solutions and incubate on ice for five minutes.

5. Add 0.2 ml of ice-cold 5 M KAc (pH 4.8). Slowly invert the tubes two times to mix the solutions and then incubate on ice for five minutes.

6. Centrifuge the lysates at maximum speed for three minutes in a minufuge. Remove the supernatant, about 0.5 ml, and transfer it to new 1.5-ml minifuge tubes.

7. Add 0.5 ml of isopropanol and precipitate the DNA (and RNA) at room temperature for 20 minutes.

8. Centrifuge the precipitated nucleic acids in the minifuge for five minutes at maximum speed. Aspirate the alcohol supernatant and dry the nucleic acid pellet as you have done in previous periods.

9. Resuspend the nucleic acid pellets in 70 μl TE buffer (pH 7.4).

10. Transfer 20 μl from each sample into separate 0.5-ml minifuge tubes. Add 3 μl of sample solution to each tube and put the mixtures into separate wells in a 1.0% agarose minigel (see the protocol for Period 7). Put λ DNA cleaved with *Eco*RI in the extra lanes of the gel for size markers. For a 12-lane gel, a good arrangement of the samples is the following: lanes 1–3, DNA from (a); lanes 4 and 9, λ DNA size markers, lanes 5–8, DNA from (b); lanes 10–12, DNA from (c). The remaining 50 μl of the DNAs from (c) will be used in Part C of this laboratory period (see below).

11. Electrophorese the samples at 150 V/gel for 90–120 minutes.

12. Stain your gel with the ethidium bromide staining solution and photograph.

C. Physical Analysis—Orientation of *nrd* Genes

1. You may want to start Part C at the same time you start Part B.

2. Select six additional Amp^{r}Tet^{r}Cap^{s}Hyu^r cultures from the overnight cultures of Period 9 (if possible).

3. Repeat Steps B2–B8 for these cultures.

4. Resuspend the nucleic acid pellets in 50 μl of TE buffer (pH 7.4).

5. Add 5 μl of 2 M NaAc (pH 6.5) to each of the DNA samples prepared from the Hyu^r cells (10 samples—four from Part B and six from Part C). Add 55 μl of isopropanol and precipitate the nucleic acids at room temperature for 20 minutes.

6. Centrifuge the precipitated nucleic acids at maximum speed for five minutes in the minifuge. Remove the alcohol by aspiration and dry the pellets as you did above.

7. Resuspend the nucleic acid pellets, which may be invisible, in 34 μl of water. If the *Bam*HI endonuclease is not at 10 units/μl, you will have to make an adjustment in the water volume (see Step 8 below).

8. Prepare the following restriction reaction mixtures for each of the 10 plasmid DNA samples:

34 μl Plasmid DNA

4 μl 10$\times$ Bam buffer (pH 8.0)

1 μl *Bam*HI (10 units/μl)

1 μl RNase T1 (see Note 1)
_____

40 μl

Incubate at 37°C for one hour.

9. Heat the reaction mixtures to 75°C for five minutes to inactivate the *Bam*HI endonuclease.

10. Freeze the reactions at −20°C until Period 11.

IV. SUMMARY

A. Evaluate the Biological Assays of Period 9

B. Size Analysis

Centrifuge overnight cultures	
Resuspend	0.1 ml GTE buffer (pH 7.5) with lysozyme
Incubate five minutes at room temperature	
Add	0.2 ml lysis solution
Incubate five minutes on ice	
Add	0.2 ml 5 M KAc (pH 4.8)
Incubate five minutes on ice	
Centrifuge three minutes	
Recover supernatants; add	0.5 ml isopropanol
Incubate 20 minutes at room temperature	
Centrifuge five minutes	
Aspirate alcohol and dry pellets	
Resuspend pellets	70 μl TE buffer (pH 7.4)
Electrophorese on 1% agarose gel	20 μl

C. Orientation Analysis

Centrifuge overnight cultures	
Resuspend	0.1 ml GTE buffer (pH 7.5) with lysozyme

Incubate five minutes at room temperature

Add   0.2 ml lysis solution

Incubate five minutes on ice

Add   0.2 ml 5 M KAc (pH 4.8)

Incubate five minutes on ice

Centrifuge three minutes

Recover supernatants; add   0.5 ml isopropanol

Incubate 20 minutes at room temperature

Centrifuge five minutes

Aspirate alcohol and dry pellets

Resuspend pellets   50 μl TE buffer (pH 7.4)

Add (to *10* DNA samples):   5 μl 2 M NaAc (pH 6.5)
                                     55 μl isopropanol

Incubate 20 minutes at room temperature

Centrifuge five minutes

Aspirate alcohol and dry pellets

Resuspend DNA: 34 μl H$_2$O
                                4 μl 10X Bam buffer (pH 8.0)
                                1 μl *Bam*HI (10 units/μl)
                                1 μl RNase
                                _____
                                40 μl

Incubate one hour at 37°C

Heat five minutes at 75°C

Freeze samples at −20°C

V. NOTES AND QUESTIONS

1. The RNase solution is necessary to ensure that small restriction fragments are not obscured by ribosomal RNA, which constitutes about 95% of the nucleic acids in the samples. The RNase, at a concentration of 1000 units/ml is boiled for five minutes in 0.1 M NaAc (pH 5.0), 1 mM EDTA to inactivate possibly contaminating DNases.

2. Compare the transformation frequencies you obtained with those of the class. Explain any discrepancies. Use the class results to answer the following questions.

3. From the transformation frequencies, what is the ratio of plasmids with inserts to those lacking inserts? Is this value to be expected from the calculations of Period 8, Note 3?

4. What is the ratio of recombinant plasmids with *nrd*⁺ genes to recombinant plasmids lacking these genes? Is this ratio expected? Explain.

VI. INTERIM BETWEEN PERIODS 10 AND 11

The instructor will start a 5 ml RB (3-1) culture of JM 107 from a single colony taken from a supplemented minimal medium plate the day before class. Two hours before class, the culture will be diluted 50-fold in RB.

Orientation Analysis of Recombinant Plasmids and Titration of Phage M13

I. INTRODUCTION

In this period, you will finish determining the orientation of the nrd^+ inserts. You should prepare maps of the two simpler pBRnrd recombinant molecules that could form and determine the positions of the *Bam*HI restriction sites (use Figures 3-2 and 4-1 on pages 22 and 26). With these maps you can predict the sizes of the restriction fragments that would arise for both orientations. If both orientations are found, then the *Eco*RI fragment containing the nrd^+ genes is self-sufficient for expression.

For Period 12, two things must be done: (1) a culture of JF427, transformed with one of the recombinant plasmids that has been characterized with respect to genetic markers, size, and orientation should be started—the plasmid in this culture will be used as a source of nrd genes for subcloning into phage M13; (2) M13 stocks should be titered—their concentrations should be 10^{11}–10^{12} phage/ml.

Strain JM107 does not contain a functional β-galactosidase gene. Both M13mp10 and M13mp11 contain a fragment of this gene that can restore enzymatic activity even though the base sequence has been modified by the introduction of cloning sites (see Chapter 5). The slightly modified enzyme is biologically active until a segment of DNA is inserted into any one of the cloning sites; in that case a *fusion polypeptide* containing an amino acid sequence specified by the recombined segments is formed. By including the substrate 5-bromo-4-chloro-3-indolyl-β-D-galactoside (X*gal*) in the agar, the presence of a functional β-galactosidase gene fragment in M13 can be detected by observing an M13 plaque, because X*gal*, which is colorless, is cleaved by an active enzyme to produce galactose and a blue product. An M13 phage lacking inserted sequences produces a blue plaque, but one containing an inserted DNA sequence does not produce active enzyme so its plaques are colorless. All M13 used in this period should be free of inserts so the plaques will be blue.

II. MATERIALS PER TEAM

1. *Bam*HI-cleaved DNA samples from Period 10

2. 50 μl of STE-saturated phenol (1-7)

3. 50 ml of 20× TAE buffer (pH 7.2) (1-14); use 1.25 ml if recycled reservoir buffer is used

4. 0.3 g of agarose (2-4)

5. 30 μl of sample solution (1-15)

6. Ethidium bromide staining solution (1-16)

7. A growing culture of JM107 (4-7)

8. Stocks of M13mp10 (4-8) and M13mp11 (4-9)

9. 10 μl of 0.1 M IPTG (1-23)

10. 50 μl of 2% X*gal* (1-24)

11. 3 ml of RA soft agar (3-2/s)

12. One RA agar plate (3-2)

13. 40 ml of 0.9% NaCl (1-25)

14. 1 ml of RB medium (3-1)

15. Twelve 0.5-ml minifuge tubes

16. Six 25-ml dilution tubes

17. 1 μl of 50 mg/ml ampicillin (1-26)

III. PROCEDURES

A. Orientation Analysis of *nrd* Genes in pBR*nrd*

1. Thaw the DNA solutions prepared in Period 10.

2. Add 50 μl of STE-saturated phenol to each sample. Vortex for 30 seconds and then centrifuge for one minute in a minifuge.

3. Remove a 50 μl aliquot from each upper (aqueous) phase and place it in a separate minufuge tube. Discard the phenol layers.

4. Add 10 μl of sample buffer.

5. Using standard procedures (see Period 7), electrophorese 10–20 μl of each sample in a 1.0% agarose gel to separate the *Bam*HI-cleavage fragments. Add λ DNA cleaved with *Eco*RI to lanes 4 and 9 for size references.

6. Analyze the size of the restriction fragments and determine the orientations of the inserted *nrd* segments.

7. Compare your data with that obtained by other teams.

B. Titration of M13

1. Obtain a culture of JM107 whose cell density is 3–5 $\times$ 10^8 cells/ml.

2. Add 0.2 ml of JM107, 10 μl of 0.1 M IPTG (an inducer of the *lac* operon), and 50 μl 2% X*gal* to 3 ml of RA soft agar. Pour on one plate to provide a lawn for spot tests with M13.

3. Prepare a dilution series in 15-ml tubes in which 50 μl of M13 is successively diluted four times into 4.95 ml of 0.9% NaCl. Then, do two successive dilutions of 0.1 ml to 0.9 ml of 0.9% NaCl. This will give dilutions of 10^2, 10^4, 10^6, 10^8, 10^9, and 10^{10}. In each case, 10 μl will be plated instead of the usual 0.1 ml using the procedure in the next step. The dilution series is shown in Figure P11-1.

4. Label six positions on the underside of the agar plate prepared in Step 2. Then, deposit 10 μl from each of the dilution tubes except the first (the 100-fold dilution) in six different positions on this plate. Before the droplets are absorbed, tilt the plate slightly and rotate it to spread each droplet to a diameter of about 1 cm, as shown in Figure P11-1.

5. Incubate the plate overnight at 37° C.

C. Overnight Culture of JF427/pBR*nrd*

Prepare an overnight culture in 1 ml of RB medium supplemented with ampicillin at 50 μg/ml, using cells containing one of the pBR*nrd* recombinant plasmids.

Figure P11-1 Minitration procedure for phage M13.

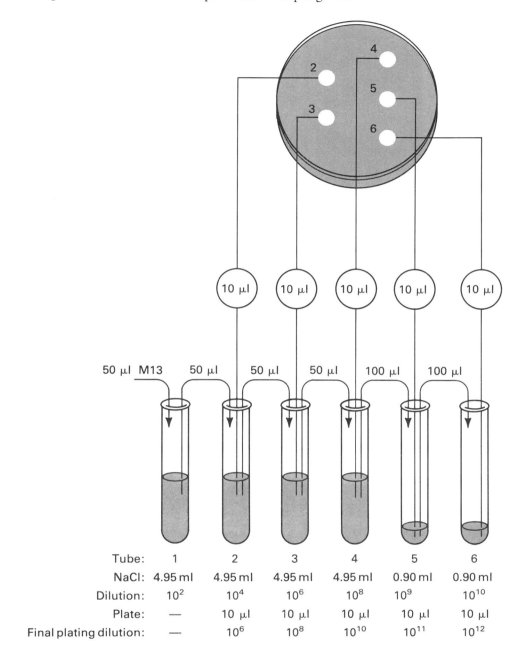

Tube:	1	2	3	4	5	6
NaCl:	4.95 ml	4.95 ml	4.95 ml	4.95 ml	0.90 ml	0.90 ml
Dilution:	10^2	10^4	10^6	10^8	10^9	10^{10}
Plate:	—	10 μl	10 μl	10 μl	10 μl	10 μl
Final plating dilution:	—	10^6	10^8	10^{10}	10^{11}	10^{12}

IV. SUMMARY

A. Orientation Analysis

DNA from Period 10	50 μl
Add	50 μl STE-saturated phenol
Vortex 30 seconds, centrifuge one minute	
Recover supernatant	
Electrophorese	20μl

B. Titration of M13

	3.0 ml RA soft agar
Add	0.2 ml JM107
Add	10 μl 100 mM IPTG
Add	50 μl 2% X*gal*
Spread onto RA plate	
To	4.95 ml 0.9% NaCl
Add	0.05 ml M13 stock
Vortex	5.00 ml
Remove	0.05 ml
Add	4.95 ml 0.9% NaCl
Vortex	5.00 ml
Plate	10 μl = 10^6 dilution
Remove	0.05 ml
Add	4.95 ml 0.9% NaCl
Vortex	5.00 ml
Plate	10 μl = 10^8 dilution
Remove	0.05 ml
Add	4.95 ml 0.9% NaCl
Vortex	5.00 ml
Plate	10 μl = 10^{10} dilution
Remove	0.5 ml
Add	4.5 ml 0.9% NaCl
Vortex	5.0 ml
Plate	10 μl = 10^{11} dilution
Remove	0.5 ml
Add	4.5 ml 0.9% NaCl
Vortex	5.0 ml
Plate	10 μl = 10^{12} dilution

C. Inoculate 1-ml Overnight Culture with JF427/pBR*nrd*

V. NOTES AND QUESTIONS

1. What are the sizes of the fragments produced by *Bam*HI digestion of pBR*nrd* in both of the possible orientations?

2. What ratio of the orientations did you obtain in your experiment? How do these results compare with that of the entire class?

3. How do you explain your results?

VI. INTERIM BETWEEN PERIODS 11 AND 12

The instructor will remove the agar plates with the M13 plates from the 37°C incubator and will put them at 4°C to prevent further growth.

Amplification of Recombinant Plasmids and Cleavage of M13 DNA with Restriction Enzymes

I. INTRODUCTION

In this period, you will prepare the DNA needed for subcloning the *nrd* genes from pBR*nrd* into M13. The recombinant plasmid will be amplified as was done for pBR325 in Period 5. You will be provided with the replicating form (RF) of M13 DNA (a double-stranded molecule) since the procedure needed for isolating it from infected cells is the same as that used earlier in the course for isolation of plasmid DNA. The M13 DNA will be cleaved with the restriction enzymes *Bam*HI and *Sst*I, which together cut out a very small piece of M13 DNA (see Chaper 5) leaving a cloning site having different termini. Thus, all DNA molecules that you wish to clone into M13 will need one *Sst*I terminus and one *Bam*HI terminus. This requirement specifies a particular orientation of the inserted DNA into the vector; such a protocol is called *forced cloning*.

During this laboratory period you will be given an introductory lecture on the use of M13 as a cloning vector.

II. MATERIALS PER TEAM

1. Overnight culture of JF427/pBR*nrd* (from Period 11)

2. 35 ml of RB medium (3-1)

3. 0.3 ml of chloramphenicol (17 mg/ml in ethanol), freshly made (1-27)

4. One 125-ml Erlenmeyer flask

5. Several 5-ml tubes for collecting culture samples

6. Stocks of M13mp10 RF (4-8a) and M13mp11 RF (4-9a), one set per class

7. 10 μl of 10× *Eco*RI buffer (pH 7.2) (1-13)

8. 20 μl of *Sst*I (1 unit/μl) (4-15)

9. 2 μl of *Bam*HI (10 units/μl) (4-14)

10. 20 μl of 0.09 M EDTA (pH 8.3) (1-28)

11. 120 μl of STE-saturated phenol (1-7)

12. 14 μl of 2 M NaAc (pH 6.5) (1-8)

13. 200 μl of isopropanol (2-2)

14. Four 0.5-ml minifuge tubes

III. PROCEDURES

A. Plaque Counting

Count the plaques on your M13 titer plate and compare your result to that of the entire class.

B. Amplification and Preparation of Plasmid DNA

1. Measure the absorbance of the overnight culture of JF427/pBR*nrd*.

2. Put 35 ml of RB medium in a 125-ml flask and add enough of the overnight culture to yield an absorbance of 0.05.

3. Follow the procedure for plasmid amplification (Period 5) but substitute chloramphenicol at 170 μg/ml for spectinomycin (add 0.35 ml of the concentrated stock solution of chloramphenicol). (Note that chloramphenicol is used since the insertion of the *nrd* gene has inactivated the *cap* gene of pBR325.)

C. Digestion of M13 RF with *Sst*I and *Bam*HI

1. Obtain M13mp10 RF and M13mp11 RF from the instructor. The concentration of each will be 100 μg/ml.

2. Using 1 μg of each RF, cleave with *Sst*I, utilizing the same conditions used in Period 7 for digestion with *Eco*RI. For both samples, carry out the reaction in a 0.5-ml minifuge tube. The reaction mixture will consist of:

10 μl RF

5 μl 10× *Eco*RI buffer (pH 7.2)

10 μl *Sst*I (10 units)

25 μl Water

———

50 μl

Incubate at 37°C for 30 minutes.

3. Add 1 μl (10 units) of *Bam*HI (see Note 1), and continue the incubation at 37°C for another 30 minutes.

4. Stop the reaction by adding 10 μl of 0.09 M EDTA (pH 8.3).

5. Deproteinize the reaction mixture by adding 60 μl of STE-saturated phenol and vortexing for 30 seconds. Then, centrifuge for one minute in the minifuge.

6. Remove 60 μl of each of the aqueous phases and add each to a new minifuge tube. Add 7 μl of 2 M NaAc (pH 6.5) and 100 μl of isopropanol.

7. Store the cleaved DNA samples at −20°C until Period 14.

IV. SUMMARY

A. Plaque Counting

B. Amplification of Recombinant Plasmid pBR*nrd*

Dilute overnight JF427/pBR*nrd* to 35 ml (3–5 × 10^7 cells/ml)

Incubate at 37°C

Plot growth

Add 0.35 ml chloramphenicol (17 mg/ml)

Continue incubation at 37°C overnight

Provides plasmid for Period 13

C. Digestion of M13mp10 RF and M13mp11 RF with *Sst*I and *Bam*HI:

$$10\,\mu l \quad RF$$
$$5\,\mu l \quad 10\times EcoRI \text{ buffer (pH 7.2)}$$
$$\underline{\quad}\,\mu l \quad \text{Water}$$
$$\underline{\quad}\,\mu l \quad SstI \text{ (5-10 units)}$$
$$\overline{50\,\mu l}$$

Incubate the reaction 30 minutes at 37°C

Add $\qquad$ 1 μl *Bam*HI (5–10 units)

Incubate the reaction 30 minutes at 37°C

Add $\qquad$ 10 μl 0.09 M EDTA (pH 8.3)

Add $\qquad$ 60 μl STE-saturated phenol

Vortex 30 seconds, centrifuge one minute

Remove the aqueous phase, about 60 μl aqueous phase

Add $\qquad$ 7 μl 2 M NaAc (pH 6.5)

Add $\qquad$ 100 μl isopropanol

Store cleaved DNAs in ethanol at 20°C until Period 14.

V. NOTES AND QUESTIONS

1. The conditions that maximize the activity of *Bam*HI and *Sst*I are not the same as for *Eco*RI. However, the *Eco*RI buffer is sufficient for these experiments but, to be sure that digestion is complete, an excess of enzyme is added (see Appendix F).

2. What is the titer of your phage stock? Compare your results with those of the other teams. All results should be similar.

VI. INTERIM BETWEEN PERIODS 12 AND 13

After 24 hours of growth, the instructor will place the cells, with the amplified plasmids at 4°C until Period 13.

Preparation of Recombinant Plasmid pBR*nrd* DNA

I. INTRODUCTION

In this period, isolation of plasmid DNA will again be carried out as in Period 6, except that you will not take the time to remove the RNA. The contaminating RNA that remains when this procedure is used will not interfere with any of the enzymatic reactions but will make it impossible to determine the DNA concentration by an absorbance measurement. Thus, the concentration of the DNA will be determined by separating the RNA from the plasmid DNA by electrophoresis through agarose and comparing the intensity of the fluorescence of the pBR*nrd* band with bands of known amounts of pBR325 DNA. This comparison will not be particularly accurate but will be adequate for your purposes.

II. MATERIALS PER TEAM

1. An amplified culture of JF427/pBR*nrd* from Period 12

2. 0.4 ml of GTE buffer (pH 7.5) (1-9) containing lysozyme (4-18) at 1 mg/ml

3. 0.8 ml of lysis solution (1-10)

4. 0.6 ml of 5 M KAc buffer (pH 4.8) (1-11)

5. 10 ml of isopropanol (2-2)

6. 50 ml of 20× TAE buffer (pH 7.2) (1-14)—use 1.25 ml if recycled reservoir buffer is used

7. 0.3 g agarose (2-4)

8. 20 μl of sample solution (1-15)

9. One 15-ml Corex centrifuge tube

10. Ten 1.5-ml minifuge tubes

11. Ethidium bromide staining solution (1-16)

12. pBR325 from Period 6

III. PROCEDURES

1. Obtain your chloramphenicol-treated JF427/pBR*nrd* culture and transfer it in a 30- or 50-ml centrifuge tube.

2. Centrifuge the cells at 12,000 × g (10,000 rpm in a Sorvall SS-34) for five minutes at 4°C.

3. Resuspend the cell pellet in 0.4 ml of GTE buffer (pH 7.5) containing 0.4 mg of lysozyme (add the lysozyme just before use). Transfer the cells to a 15-ml Corex tube. Maintain at room temperature for five minutes.

4. Add 0.8 ml of lysis solution. Mix gently and incubate for five minutes on ice.

5. Add 0.6 ml of ice-cold 5 M KAc buffer (pH 4.8) and mix gently; incubate five minutes on ice.

6. Centrifuge the lysate for 10 minutes at 27,000 × g (15,000 rpm in a Sorvall SS-34 rotor) at 4°C or at room temperature.

7. Remove the supernatant with a Pasteur pipette and place it in a 15-ml Corex tube. Add one volume (about 2 ml) of isopropanol at 20°C.

8. Allow the nucleic acids to precipitate at room temperature for 20 minutes.

9. Centrifuge the solution at 12,000 × *g* (10,000 rpm in a Sorvall SS-34 rotor) for 10 minutes at 4°C.

10. Pour off the isopropanol, blot the rim of the tube, and vacuum dry for 15 minutes.

11. Resuspend the nucleic acid pellet in 100 μl of water and transfer it to a 0.5-ml minifuge tube.

12. Remove 5 μl and dilute it into 1 ml of water. Store the remainder at 0°C until Step 14. Measure the absorbance at both 260 nm and 280 nm and calculate the nucleic acid concentration. A protein-free sample should have a value of A_{260}/A_{280} of 1.9–2.0 (the ratio will be less than 1.9 if the sample is contaminated with protein). Using the fact that A_{260} = 1.0 corresponds to 45 and 50 μg/ml of RNA and DNA, respectively, and that the composition of the nucleic acid mixture is approximately 98% RNA and 2% DNA, estimate the concentration of the DNA in the sample.

13. Prepare an 0.8% agarose gel in TAE buffer (see Period 7).

14. Prepare four samples of pBR*nrd* DNA containing the following amounts of plasmid DNA estimated from Step 12: 0.05, 0.2, 0.5, and 1 μg—each in a total volume of 18 μl; water should be used for all dilutions. Then, add 2 μl of sample buffer to each sample. Freeze the remaining pBR*nrd* until Period 14.

15. Prepare four samples of pBR325 (from Period 7) containing the following measured amounts of DNA: 0.05, 0.2, 0.5, and 1 μg, in a total volume of 18 μl. Then, proceed as in Step 14.

16. Electrophorese the samples at 100–150 V/gel until the blue dye is 1–2 cm from the bottom of the gel. Stain and photograph the gel, as done in Period 7.

IV. SUMMARY

Resuspend pellet of chloramphenicol-treated cells	0.4 ml GTE buffer (pH 7.5) with lysozyme
Add	0.8 ml lysis buffer
Incubate five minutes on ice	
Add	0.6 ice-cold 5 M KAc (pH 4.8)
Incubate five minutes on ice	
Centrifuge 27,000 × *g,* 10 minutes	
Precipitate supernatant	
Add	2.0 ml isopropanol

Incubate 20 minutes at room temperature

Centrifuge precipitate 12,000 × g, 10 minutes

Resuspend pellet   0.1 ml H_2O

Dilute   5 μl into 1 ml H_2O

Measure absorbance at 260 and 280 nm

Prepare   0.8% agarose gel in 20× TAE buffer (pH 7.2)

Electrophorese 0.05, 0.2, 0.5, and 1 μg of pBR*nrd* DNA and the equivalent amounts of pBR325 DNA

Freeze the residual volume of your pBR*nrd* DNA

V. NOTES AND QUESTIONS

1. Compare the bands you find in your preparation of pBR*nrd* with those you found in your analysis of pBR325 in Period 7. You may need to refer to the identification guide given in the notes section of Period 6. You should find the following differences: (a) the supercoiled monomer is the dominant band in this sample; (b) the relaxed circular monomers and dimers are minor bands in this preparation; (c) a very large amount of RNA, having a large range in size, is present.

2. Using the intensities of the fluorescence of the pBR325 DNA standards, estimate the amount of pBR*nrd* in your samples.

3. Calculate the concentration of pBR*nrd* DNA in your stock sample (Step 11).

Subcloning of *nrd* $^+$ Genes into M13

I. INTRODUCTION

In this period you will join the (*Sst*I-*Bgl*II)-terminated fragments obtained from digestion of pBR*nrd* to M13 RF that has been cleaved by *Sst*I and *Bam*HI. Figure P14-1 illustrates how this protocol produces a unique orientation of the inserted fragments. However, despite the specificity imposed by the termini of the fragments, a variety of functional molecules can still form. You should draw a restriction map of pBR*nrd* and determine the structures that might be present in an annealed mixture.

II. MATERIALS PER TEAM

1. M13mp10 RF cleaved with *Sst*I and *Bam*HI from Period 12

2. M13mp11 RF cleaved with *Sst*I and *Bam*HI from Period 12

3. pBR*nrd* DNA from Period 13

4. 150 μl of low-Tris buffer (pH 7.5) (1-29)

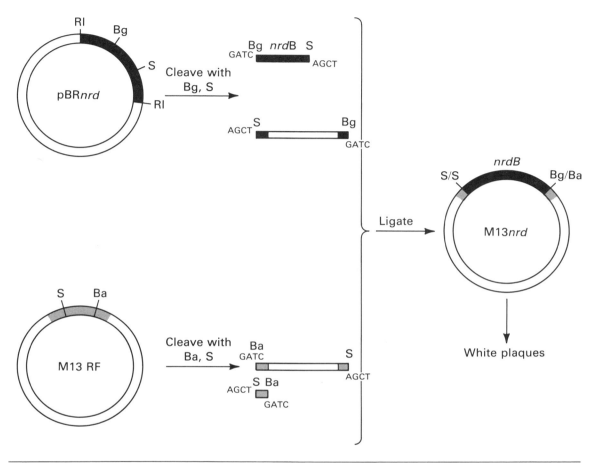

Figure P14-1 Forced cloning of the (*Bgl*II-*Sst*I)-terminated pBR*nrd* fragments into M13 RF cleaved with *Sst*I and *Bam*HI. As explained in Chapter 5, the overlapping ends of *Bgl*II-cleaved fragments and *Bam*HI-cleaved fragments are complementary even though the enzyme recognition sites are different. Ba, *Bam*HI site; Bg, *Bgl*II site; RI, *Eco*RI site; S, *Sst*I site.

5. 5 μl of 10× *Eco*RI buffer (pH 7.2) (1-13)

6. 5 μl of *Sst*I (4-15)

7. 10 μl of *Bgl*II (4-16)

8. 10 μl of 0.09 M EDTA (pH 8.3) (1-28)

9. 60 μl of STE-saturated phenol (1-7)

10. 60 μl chloroform (2-1)

11. 7 μl of 2 M NaAc (pH 6.5) (1-8)

12. 1 ml of isopropanol (2-2)

13. 20 μl of 1 mM ATP (1-17)

14. 20 μl of 10X ligation buffer (pH 7.2) (1-18)

15. 8 μl (8 units) T4 ligase (4-13)

16. Eight 1.5-ml minifuge tubes

17. 10 μl TE buffer (pH 7.4) (1-12)

III. PROCEDURES

1. Centrifuge the cleaved M13mp10 RF and M13mp11 RF DNA samples in minifuge tubes for five minutes at maximum speed. Remove the isopropanol by aspiration and dry the pellets in a vacuum for 10 minutes.

2. Resuspend the pellets in 50 μl of low-Tris buffer (pH 7.5) to make a solution having a DNA concentration of 20 μg/ml. These DNAs will be used in Step 10.

3. Cleave the pBR*nrd* DNA with both *Sst*I and *Bgl*II in the following reaction in a 0.5-ml minifuge tube.

    ___ μl   pBR*nrd* DNA (2 μg)
    5 μl   10X *Eco*RI buffer (pH 7.2)
    5 μl   *Sst*I endonuclease (1 unit/μl)
    10 μl   *Bgl*II endonuclease (1 unit/μl)
    ___ μl   H$_2$O
    ‾‾‾‾‾
    50 μl

Incubate with both enzymes acting simultaneously at 37°C for 45 minutes.

4. Stop the reaction by adding 10 μl of 0.09 M EDTA (pH 8.3).

5. Deproteinize the reaction mixture by adding 60 μl of STE-saturated phenol and vortexing for 30 seconds. Then, centrifuge for one minute in the minifuge. Reextract the aqueous phase with 60 μl chloroform.

6. Remove 60 μl of the aqueous phase. Add 7 μl of 2 M NaAc (pH 6.5) and 0.1 ml of isopropanol and precipitate at −20°C for 20 minutes.

7. Centrifuge the solution for 5 minutes.

8. Pour off the isopropanol, blot dry the rim of the tube, and vacuum dry the pellets for 15 minutes.

9. Resuspend the pellets in 80 μl of low-Tris buffer (pH 7.5); the DNA concentration should be about 20 μg/ml.

10. Prepare the following four ligation mixtures:

(L1): M13mp10 RF alone;
(L2): M13mp11 RF alone;
(L3): M13mp10 RF with pBR*nrd*;
(L4): M13mp11 RF with pBR*nrd*.

The reaction mixtures should contain the following ingredients:

5 μl M13 RF DNA cleaved with *Sst*I and *Bam*HI
5 μl pBR*nrd*, cleaved with *Sst*I and *Bgl*II, or TE buffer (1-12)
4 μl 10× ligation buffer (pH 7.2)
5 μl 1 mM ATP
2 μl T4 ligase (1 unit/μl)
$\underline{19\ \mu l}$ H$_2$O
40 μl

Incubate at 15°C overnight.

IV. SUMMARY

Precipitate M13 RF DNAs from Period 12	
Resuspend	50 μl low-Tris buffer (pH 7.5)
Digest pBR*nrd* with *Sst*I and *Bgl*II:	__ μl DNA (2 μg)
	5 μl 10× *Eco*RI buffer (pH 7.2)
	5 μl *Sst*I (1 unit/μl)
	10 μl *Bgl*II (1 unit/μl)
	$\underline{\quad \mu l}$ H$_2$O
	50 μl
Incubate 45 minutes at 37°C	
Add:	10 μl 0.09 M EDTA (pH 8.3)
	60 μl STE-saturated phenol
Vortex 30 seconds	
Centrifuge one minute	
Remove	60 μl aqueous solution
Reextract	60 μl chloroform
Add to aqueous supernatant:	7 μl 2 M NaAc (pH 6.5)
	100 μl isopropanol
Precipitate at −20°C, 20 minutes	
Centrifuge 16,000 × g for 10 minutes at 4°C	
Resuspend DNA fragments	80 μl low-Tris buffer (pH 7.5)

Prepare ligation reactions:	Vector	Plasmid/TE buffer
(L1):	M13mp10 RF	TE buffer
(L2):	M13mp11 RF	TE buffer
(L3):	M13mp10 RF	pBR*nrd*
(L4):	M13mp11 RF	pBR*nrd*

Reactions
5 μl Vector DNA
5 μl Plasmid DNA
4 μl 10X ligation buffer (pH 7.2)
5 μl 1 mM ATP
2 μl T4 ligase (1 unit/μl)
19 μl H$_2$O
40 μl

Incubate overnight at 15°C.

VI. INTERIM BETWEEN PERIODS 14 AND 15

1. The instructor will remove the ligation reactions from the 15°C water bath and freeze them.

2. The instructor will start an overnight culture of JM107 in RB. Two hours before the next class, the instructor will dilute the culture of JM107 50-fold to a total volume of 300 ml.

Period 15

Transfection with M13 Recombinant DNA

I. INTRODUCTION

In this period, cells will be rendered competent and transfected with M13 recombinant DNA in order to obtain recombinant phage.

The procedure for obtaining cells competent for transfection with M13 is similar to the procedure for plasmid transformation used in Period 8. However, the treatments following uptake of the DNA differ. When plasmids are used, a recovery growth period is needed prior to plating in order that the transformed cells can begin manufacturing products for protection against antibiotics; this step is not necessary when a plaque assay is being used. A second culture of indicator cells is also needed because the $CaCl_2$-treated cells form a poor bacterial lawn when cells are plated in soft agar and plaques would be very difficult to see. However, since the transfected cells produce phage M13, a second layer of untreated cells may be used to form a lawn. Usually, a small portion of the cells is set aside and not treated with $CaCl_2$. Because of the simpler plating procedure, transfection with M13 is faster than transformation with plasmid DNA. By using actively growing indicator cells, plaques form in four hours and the color produced by hydrolysis of the X*gal* appears in six hours.

131

II. MATERIALS PER TEAM

1. Culture of JM107

2. 25 ml of 0.05 M $CaCl_2$ (1-19)

3. 50 μl of 0.1 M IPTG (1-23)

4. 250 μl of 2% X*gal* (1-24)

5. 20 ml of RA soft agar (3-2/s)

6. Five RA agar plates (3-2)

7. Ligation reactions L1, L2, L3, and L4 from Period 14

8. 10 μl of low-Tris buffer (pH 7.5) (1-29)

9. Two 40-ml polypropylene centrifuge tubes

III. PROCEDURES

A. Preparation of Competent JM107 Cells

1. Transfer 40 ml exponentially growing culture of JM107 at $A_{600} \cong$ 0.6 into a centrifuge tube and centrifuge at $8000 \times g$ (8000 rpm in a Sorvall SS-34 rotor) for five minutes at $0°C$. Discard the supernatant and resuspend the cells in 20 ml of iced 0.05 M $CaCl_2$. After 20 minutes (still on ice) centrifuge again, and resuspend the cells in 4 ml of 0.05 M $CaCl_2$. Keep the cells at $0°C$ at all times.

2. Take the few drops of the culture remaining in the flask, add to 3 ml of fresh RB medium, and grow the cells again at $37°C$. These cells will be used to form a lawn for plating phage.

B. Transfection with M13 Recombinant DNA from Period 14

1. Add 0.2 ml of the competent cells to 10 μl of L1, L2, L3, and L4 (in four tubes) and as a control, to 10 μl of low-Tris buffer (pH 7.5) in another tube. Keep the five mixtures at $0°C$ for 45 minutes.

2. Heat the five tubes for two minutes at $42°C$ and then transfer them to room temperature.

3. To each tube add: 200 μl regrown JM107 culture (Step A2)

                                         10 μl 0.1 M IPTG

                                         50 μl 2% X*gal*

                                         3 ml RA soft agar at 50° C

Mix by vortexing and pour onto RA plates.

4. Keep the plates at room temperature on the laboratory bench for 10 minutes to let the liquid agar solidify and then incubate overnight at 37° C.

IV. SUMMARY

A. Preparation of Competent JM107 Cells

Centrifuge JM107 cells at 8000 × g for five minutes

Resuspend the cell pellet   20 ml 0.05 M CaCl$_2$

Centrifuge at 8000 × g for five minutes

Resuspend the cell pellet   4 ml 0.05 M CaCl$_2$

B. Transfection with Recombinant Phage M13

To (L1), (L2), (L3), (L4), and
low Tris buffer (pH 7.5)   10 μl each
Heat-shock 2 minutes, 42° C

Add: 0.2 ml competent JM107

                                 10 μl 0.1 M IPTG

                                 50 μl 2% X*gal*

                                 3 ml RA soft agar

Place each mixture on an RA plate

V. NOTES AND QUESTIONS

1. Consult the restriction-enzyme cleavage map for pBR*nrd* (you will need your results from Periods 10 and 11—see notes of Period 10), the map for the λ fragment containing the *nrd*$^+$ gene (Figure 3-2, page 22), pBR325 (Figure 4-1, page 26), and M13mp10 or M13mp11 (Figure 5-5, page 43). What are the possible bimolecular ligation products that can form (see Note 2 of Period 9)?

2. Which of the total bimolecular ligation products in Question 1 can circularize?

3. Which of the circularized products in Question 2 can form viable M13 recombinants?

VI. INTERIM BETWEEN PERIODS 15 AND 16

The instructor will remove the plates from the 37°C incubator and place them at 4°C until Period 16.

Period 16
Analysis of Transfected Cells

I. INTRODUCTION

The color test used in Period 15 enabled you to identify the genotypes of the M13 plaques by simple inspection of the plates. The number of plaques obtained in each experiment should be counted and classified as blue or colorless. If all went well, the control in which only buffer was used (no DNA) should not yield plaques (unless one of the solutions was contaminated with phage). L1 and L2 should yield only a few plaques and all should be blue, unless, again, the vector DNA was contaminated. The small DNA segment removed from the M13 DNA in the endonuclease treatment is rarely reinserted into the RF during the ligation reaction so the number of blue plaques mainly represents uncleaved RF molecules. L3 and L4 should have more plaques than L1 and L2 and most of these should be colorless.

To be sure that the plaque colors are correctly indicating the genotypes of the different phage types, DNA will also be analyzed. Phage DNA will be prepared from phage obtained from blue plaques in the L1 and L2 plates and from colorless plaques in the L3 and L4 plates. The L1 and L2 samples will serve as controls. Since the plaques result from slower growth of infected cells rather than from true lysis, infected cells can be taken from the plaques and placed in fresh medium. As pointed out in Chapter 5, infected cells grow very slowly and can be used as a source of phage to initiate an infection of uninfected cells, which will be added to each culture.

II. MATERIALS PER TEAM

1. 40 ml of RB medium (3-1)

2. Twelve 30-ml test tubes (preferably with screw caps)

3. Sterile toothpicks in a foil-covered beaker

4. One plate of JM107 from Period 10

III. PROCEDURES

1. Inoculate 40 ml of RB medium with a loopful of JM107 cells from the minimal glucose plate and grow for one hour at 37° C.

2. Pipette 2 ml of growing cells into 12 × 30-ml test tubes. Be sure that the tubes fit either in a roller drum or in a test tube rack mounted in a water-bath shaker. If the latter is used, place the tubes at an angle so there will be a large surface area for efficient aeration.

3. Using seven sterile toothpicks, transfer one blue plaque from each of the control plates L1 and L2 and five white plaques from plates L3 and L4. Drop the toothpicks into the individual tubes, place the tubes in a roller drum or shaker, and grow overnight at 37° C. Be sure to code each tube with the number of the plate (L1, and so on) and assign a number to each plaque.

IV. SUMMARY

Grow JM107 for one hour	40 ml RB
Dispense into 12 × 30-ml test tubes	2 ml
Add by toothpicking	5 white plaques from L3
	5 white plaques from L4
	1 blue plaque from L1
	1 blue plaque from L2
Incubate the 12 cultures overnight at 37° C	

V. NOTES AND QUESTIONS

1. What are the frequencies of blue and colorless plaques from the transformations with ligation reaction mixtures L1, L2, L3, and L4?

2. When you made pBR325/λ DNA recombinants, a significant percentage of the products were pBR325 plasmids without inserts (see Questions 5 and 6 of Period 10). In this exercise, you should have observed a very small number of M13 RF molecules without inserts in the absence (L1 and L2) or presence (L3 and L4) of pBR*nrd* DNA. Why?

3. What events can cause colorless plaques to form?

VI. INTERIM BETWEEN PERIODS 16 AND 17

The instructor will transfer the cultures from the 37°C incubator to 4°C until Period 17.

Physical Analysis of M13/pBR*nrd*$^+$ Recombinant Phages

I. INTRODUCTION

In the joining reaction between doubly digested M13 DNA and pBR*nrd* DNA (Period 14) recombinant phage DNA molecules formed that contained pBR*nrd* DNA. The color test (Period 15) indicated whether the β-galactosidase coding sequence had been interrupted, but does not tell which of the two fragments of the pBR*nrd* DNA have been inserted. A test for homology between the recombinant phage DNA molecules can provide this information (see Chapter 5). In this test, single-stranded DNA molecules obtained from M13mp10 and M13mp11 recombinant phages are hybridized. Since M13 particles contain only (+) strands, the phage sequences of the M13mp10 and M13mp11 DNA molecules cannot hybridize. However, if *nrd*$^+$ genes have been inserted in recombinant DNA molecules, the orientations of the inserted DNA in M13mp10 and in M13mp11 will be different and hence the inserted segment in one M13mp10 single strand will be complementary (with respect to base pairing) to the inserted segment in M13mp11. Thus, single strands isolated from phage particles can hybridize but only by base pairing between the inserted sequences (assuming they are complementary). Such base pairing would result in formation of a single-stranded figure-8-like molecule whose electrophoretic mobility is much less than that of either of the free single-stranded circles and which is easily detected. This test for homology is called a *C-test*.

To obtain the recombinant phage particles, cultures grown to stationary phase are centrifuged and the supernatants, which contain extruded phage, are collected. The yield of phage is so high (1000/cell) that the phage DNA concentration is sufficient for detection in agarose gels without further purification. Electrophoresis of unpurified samples is called DIGE, or *direct gel electrophoresis*.

In both experiments, the phage coat is disrupted by the addition of a detergentlike SDS.

II. MATERIALS PER TEAM

1. 12 M13 cultures from Period 16

2. 100 ml of 20× TAE Buffer (pH 7.2) (1-14)—use 2.5 ml if recycled reservoir buffer is used

3. 0.6 g of agarose (2-4)

4. 72 μl of loading buffer (pH 8.3) (1-30)

5. 20 μl of 2% SDS (1-31)

6. Ethidium bromide staining solution (1-16)

7. Twenty 0.5-ml minifuge tubes

8. 200 μl of light mineral oil (2-5)

III. PROCEDURES

A. DIGE Analysis

1. Remove 1 ml from each of the 12 overnight cultures and transfer to a 1.5-ml minifuge tube.

2. Centrifuge the cells for one minute.

3. Remove 10 μl from the supernatants of each tube, place in a new tube, and add 1 μl of 2% SDS plus 3 μl of loading buffer (pH 8.3).

4. Electrophorese the samples in a 1% agarose gel for 1.5 hours at 100–150 V/gel.

5. Stain the gel with ethidium bromide and photograph.

B. C-test

1. Place 20 µl of the supernatant of one M13mp10-recombinant culture in each of 10 tubes. To each tube add either 20 µl of one of five M13mp11-recombinant cultures or one of five M13mp10-recombinant cultures (see Figure P17-1). The second set of five tubes are the control tubes in which hybridization should not occur.

2. To each tube, add 1 µl of 2% SDS to lyse the phage.

3. To prevent evaporation, overlay each tube with 20 µl of mineral oil.

4. Incubate the tubes at 65°C for at least 90 minutes to allow hybridization to occur.

5. When the DIGE gel is completed, remove the gel, wash the plate, and pour a second 1% agarose gel in TAE buffer.

6. Add 4 µl of loading buffer (pH 8.3) to each of the C-test reaction mixtures and then load 15 µl of each of the 10 C-test tubes into the gel and electrophorese at 100–150 V/gel for 90 minutes. To obtain the samples from the C-test tubes, plunge the Pipetteman tip through the mineral oil. The mineral oil that is transferred to the gel with your samples will not interfere with the electrophoresis.

7. Stain the gel with ethidium bromide and photograph.

Figure P17-1 The C-test procedure. The clones are designated M13mp10A through M13mp10E and M13mp11A through M13mp11E. M13mp10A serves as the test sample. In addition to DNA, SDS and mineral oil are added to the reaction.

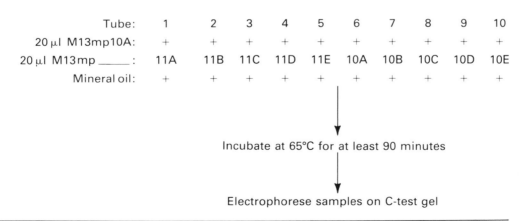

IV. SUMMARY

A. DIGE Analysis

Pellet cells

Remove supernatant

Add:

1 ml overnight cultures

10 μl sample

1 μl 2% SDS

3 μl loading buffer (pH 8.3)

Electrophorese

B. C-test Analysis (10 Assays)

10 aliquots of one M13mp10 recombinants each

20 μl

Add M13mp10 and M13mp11 recombinants

20 μl

Add:

1 μl 2% SDS

4 μl loading buffer (pH 8.3)

20 μl mineral oil

Incubate 65°C for 90 minutes or more

Electrophorese

V. NOTES AND QUESTIONS

1. Combine the results of your C-test with those of your student colleagues, and determine the frequency of cloning of each of the DNA strands of the *nrd* gene. Is the insertion frequency of the two strands equal?

2. Did your C-test hybridizations proceed to completion? What could you do to increase the proportion of hybridized to nonhybridized molecules in those mixtures that contained recombinant phage M13 with complementary inserts?

3. The bands in the stained gel containing C-test DNAs may be more diffuse than what you have seen before. This is in part due to the higher concentration of salt in the sample. Moreover, there may be an extra band at the top of the gel and a diffuse spot close to the dye marker; these bands may be chromosomal DNA and RNA, respectively, from lysed cells.

4. If you had positive C-test results, M13 phage without an obvious inserted DNA fragment, and high numbers (>1000) of blue plaques in the L1 and/or L2 ligation reactions (Period 14), propose an explanation based on the inactivity of one of the restriction enzymes.

Period 18

Site-specific Mutagenesis of the *lac* DNA in M13mp10-16

I. INTRODUCTION

A frameshift in the sequence of any gene generally leads to loss of function unless the frameshift is very near the carboxyl terminus of the protein. Frameshifts are usually created *in vivo* by growth of the organism in the presence of DNA intercalating agents. An interesting frameshift was created in the *lac* gene of M13mp10 by a technique illustrated in Figure P18-1. The *lac* gene in phage M13mp10–16 contains a base deletion plus a base substitution. The mutation was engineered by site-specific mutagenesis using an oligonucleotide primer 16 bases long. The new molecule was shortened in length by one base and therefore has a +1 frameshift, which is mutagenic. The phage carrying this frameshift is called M13mp10–16; because it is *lac⁻*, it produces a colorless plaque on an X*gal* plate.

In this period, you will use M13mp10–16 as a substrate for site-specific mutagenesis. Although in practice site-specific mutagenesis is usually carried out to generate a mutant, in this course the technique will be used to create a revertant, primarily because it is possible to complete the procedure in the time available, and a minimum of equipment and material will be used. The standard color test with X*gal* will indicate the presence of a functional alpha peptide.

In the usual procedure, a mutant sequence is used as a primer to copy the remainder of the gene (which is the template). In the experiment you will do, a restriction fragment from a plasmid that contains the wild-type

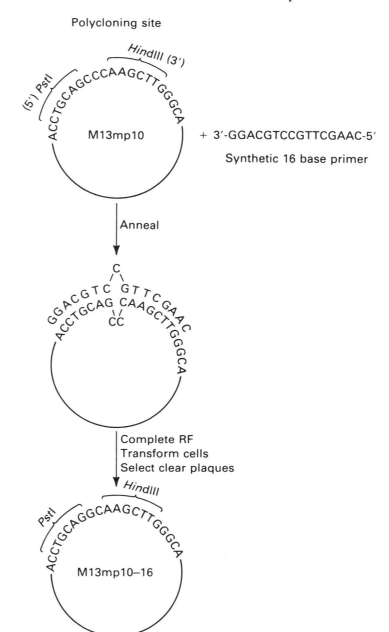

Figure P18-1 Frameshift mutation to create M13mp10–16.

sequence will be used to produce the revertant. The plasmid is pUC12, which contains a *Hae*II fragment carrying *lac* DNA (Figure P18-2). When cleaved with *Hae*II, three fragments containing 438, 370, and 1922 base pairs, respectively, are produced. Only the fragment with 438 base pairs contains the *lac* sequence and is hybridizable with M13mp10–16 DNA. In this period, M13mp10–16—isolated from phage particles—and a mixture of the *Hae*II fragments will be boiled to denature all of the DNA and the mixture will be subjected to annealing conditions (65° C). The mixture

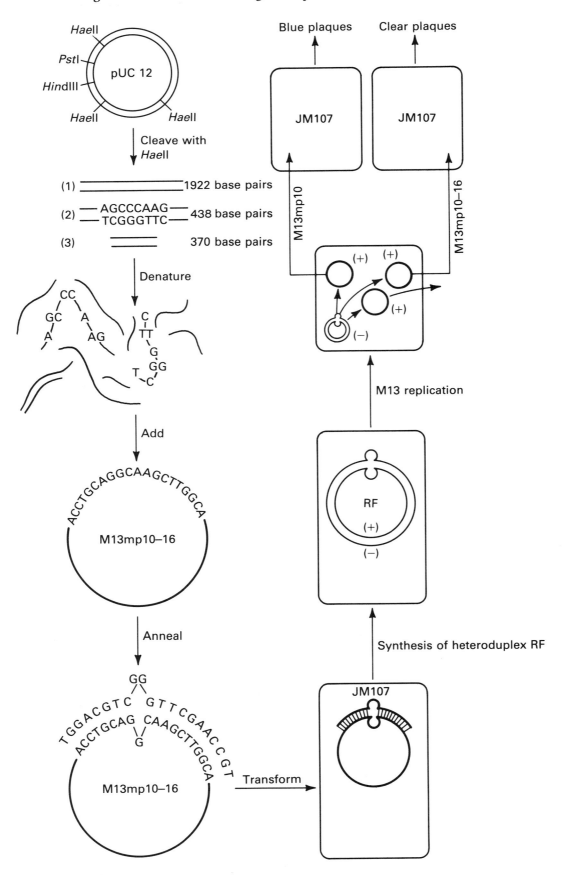

Figure P18-2 *In vitro* mutagenesis procedure.

144

will be used to transform JM107, as described in Period 17. All of the DNA will be taken up by the cells, but only structures containing intact M13 DNA will be able to replicate. The hybrid molecule, consisting of an M13mp10–16 (+) strand (the strand contained in the phage particle) and (−) strand of the 438 fragment with 438 bases will be converted to a double-stranded replicating form of M13 by bacterial DNA polymerase I; the M13mp10–16 (+) strand is the template and the fragment is the primer. Note that the intracellular (−) strand will contain the functional *lac* sequence and serve as a template for synthesis of progeny phage particles that also contain a functioning *lac* gene. Thus, these progeny phage will form blue plaques. The complete procedure is shown diagrammatically in Figure P18-2.

You will be given a lecture on the applications of site-specific mutagenesis *in vitro* during the extra time of this laboratory period.

II. MATERIALS PER TEAM

1. 0.1 µg of M13mp10–16 single-stranded DNA (4-10)

2. 2 µg of pUC12 DNA (4-11)

3. 4 µl of 10× *Hae*II buffer (pH 8.1) (1-32)

4. 2 µl of *Hae*II endonuclease (2.5 units/µl) (4-17)

5. 20 µl of 20× SSC (1-33)

6. 40 µl of mineral oil (2-5)

7. A minifuge tube

III. PROCEDURES

1. Prepare the following restriction endonuclease reactions:

	(a)	(b)
pUC12 DNA (1 µg/µl)	2 µl	—
10× *Hae*II buffer (pH 8.1)	2 µl	2 µl
*Hae*II endonuclease (2.5 units/µl)	1 µl	1 µl
H₂O	15 µl	17 µl
	20 µl	20 µl

Incubate both tubes at 37°C for 30 minutes. Cleavage of pUC12 by *Hae*II will occur in reaction (a); reaction (b) is a control (see Note 1).

2. Prepare a boiling water bath.

3. Terminate the cleavage reaction by putting the reactions in the boiling water for five minutes. Put a pinhole in the cap to keep the tubes from blowing apart.

4. Add 31 μl of water and 1 μl of M13mp 10–16 single-stranded DNA to both reactions before the boiled solution cools.

5. Add 10 μl of 20× SSC and 20 μl of mineral oil to both reactions.

6. Incubate both tubes at 65° C overnight to anneal the complementary DNAs.

IV. SUMMARY

Two reactions:	(a)	(b)
pUC12 DNA	2 μl	—
10× *Hae*II buffer (pH 8.1)	2 μl	2 μl
*Hae*II endonuclease	1 μl	1 μl
H$_2$O	15 μl	17 μl
	20 μl	20 μl

Incubate at 37° C for 30 minutes
Heat at 100° C for five minutes

Add H$_2$O	31 μl	1 μl
Add M13mp10–16DNA	1 μl	1 μl
Add 20× SSC	10 μl	10 μl
Add mineral oil	20 μl	20 μl

Incubate at 65° C overnight

V. NOTES

If you wish, you may confirm that the *Hae*II cleavage reaction occurred by electrophoresing 10% of the reaction mixtures on a 1% agarose gel, as done in Period 12. If you do this, replace the lost volume of your sample with water.

VI. INTERIM BETWEEN PERIODS 18 AND 19

1. The instructor will remove the hybridization reactions from the 65°C water bath and will place the samples in the refrigerator.

2. The instructor will grow an overnight culture of JM107 and will dilute 6 ml into 300 ml RB medium (3-1) two hours before class.

Analysis of the Site-specific Mutant Product

I. INTRODUCTION

In this period, the DNA you have prepared will be used in a transfection experiment. If everything went well, roughly 1–5% of the plaques should be blue when the reaction mixture containing pUC12 DNA is used (see notes).

II. MATERIALS PER TEAM

1. Exponentially growing culture of JM107

2. 6 ml of 0.05 M $CaCl_2$ (1-19)

3. 3 ml RB medium (3-1)

4. 30 μl of 0.1 M IPTG (1-23)

5. 150 μl of 2% Xgal (1-24)

6. 9 ml of RA soft agar (3-2/s)

7. Three RA agar plates (3-2)

8. 10 μl of M13mp10–16 DNA from Period 18

9. 10 μl of annealed DNA (M13mp10–16 + *Hae*II fragments) from Period 18

10. 10 μl of low-Tris buffer (pH 7.5)

11. Two 40-ml polypropylene centrifuge tubes

III. PROCEDURES

A. Preparation of Competent JM107 Cells

1. Transfer 10 ml of an exponentially growing culture of JM107 ($A_{600} \cong 0.6$) into a centrifuge tube and centrifuge for five minutes at 4000 $\times$ g (5000 rpm in a Sorvall SS-34 rotor) at 0° C.

2. Resuspend the cells in 5 ml of ice-cold 0.05 M $CaCl_2$. Let the cells sit for 20 minutes on ice, centrifuge again as in Step 1 and resuspend them in 1 ml of 0.05 M $CaCl_2$. After 30 minutes on ice, these are competent cells. Keep the cells on ice at all times.

3. Add a few drops of the JM107 culture to 3 ml of fresh RB medium and incubate the culture to grow cells for plating.

B. Transfection with Recombinant DNA Prepared in Period 18

1. Prepare the following three samples:

(a) Add 0.2 ml competent JM107 cells to 10 μl of reaction (a) from Period 18.

(b) Add 0.2 ml competent JM107 cells to 10 μl of reaction (b) from Period 18.

(c) Add 0.2 ml competent JM107 cells to 10 μl of low-Tris buffer (pH 7.5). This reaction will serve as a control.

2. Keep the three samples at 0° C for 30 minutes.

3. Place the three samples in a 42° C water bath for two minutes and then transfer them to room temperature.

4. To each of the samples add: 0.2 ml growing JM107
10 μl 0.1 M IPTG
50 μl 2% X*gal*
3 ml RA soft agar

5. Vortex and pour each mixture onto an RA plate. Let the plates sit for 10 minutes on your laboratory bench, and then incubate the plates overnight at 37°C.

IV. SUMMARY

A. Preparation of Competent JM107 Cells

Centrifuge at 4000 × g for five minutes	10 ml JM107 cells
Resuspend (at 0°C)	5 ml 0.5 M $CaCl_2$
Incubate 20 minutes at 0°C	
Centrifuge at 4000 × g for five minutes	
Resuspend (at 0°C)	1 ml 0.05 M $CaCl_2$
Restart JM107 culture	3 ml fresh RB medium
Incubate fresh JM107 culture at 37°C	

B. Transfection with Recombinant DNA

Three hybridization reactions:

(a) M13mp10–16 DNA plus *Hae*II primers	10 μl
(b) M13mp10–16 DNA without primers	10 μl
(c) low-Tris buffer (pH 7.5)	10 μl
To each, add:	0.2 ml growing JM107
	10 μl 0.1 M IPTG
	50 μl 2% X*gal*
	3 ml RA soft agar

Plate the three mixtures on RA plates

V. NOTES

Only 1–5% of the plaques that contain the mutagenized M13 phage will be blue because every M13 will not be annealed with the primer; synthesis of the complementary strand can occur on the single-stranded, unprimed template by initiation at the site on the phage DNA at which initiation normally occurs *in vivo* (the *ori* site).

Period 20
Conclusion

This is your last formal laboratory period and you should analyze the plates prepared in Period 19. Before you leave, you must discard all unusable materials and clean your laboratory equipment! Finally, there will be a discussion of the course.

Part Three
Appendices

Appendix A

Solutions, Reagents, and Supplies Used in this Course

1. SOLUTIONS

In this section, all of the solutions used in the 20 laboratory sessions and interim periods are listed. The amounts used are per team of students. Many of the solutions are used in sparing amounts; consequently, we have found that teams of students often prefer to divide up the labor and make common solutions for an entire class. The accuracy in preparing larger amounts of common solutions is generally greater than in preparing smaller quantities of the same solutions.

For ease in preparation, we suggest that students prepare stock quantities of the more common reagents and prepare specific solutions from these stocks. These common solutions include:

2 M Tris (pH 7.5)—(adjust pH as required in dilutions)☆

5 M NaCl

0.5 M EDTA (pH 8.0)☆

16.6 N HCl (concentrated HCl from the manufacturer)

10 N NaOH

☆Included in Chemicals Package (see page 163)

	Solutions	Amounts Required	Periods Used
(1-1)	1 M MgSO$_4$	2.2 ml	3, 4
(1-2)	Phage dilution (PD) buffer (pH 7.5) 0.01 M MgSO$_4$ 0.01 M Tris (pH 7.5)☆	125 ml	3,4
(1-3)	DNaseI (10 mg/ml)⊕ Store at −20°C	10 μl	3
(1-4)	RNase T1 (1000 units/ml)⊕ (a) Dissolve in 0.1 M NaAc (pH 5.2), 0.001 M EDTA.☆ (b) Heat in a boiling water bath for five minutes to inactivate any DNases. Store at −20°C	70 μl	3, 6, 10
(1-5)	100× spectinomycin (30 mg/ml) in H$_2$O◊ Make up fresh before use		
(1-6)	STE (to saturate phenol) 0.1 M NaCl 0.01 M Tris (pH 7.5)☆ 0.001 M EDTA☆		
(1-7)	STE-saturated phenol (a) Mix 2 volumes redistilled phenol (melted),☆ 1 volume STE (1-6). (b) Let the aqueous phase separate from the organic phase; the aqueous phase will float on the buffer-saturated phenol phase. (c) Store the phenol in a brown bottle at 4°C.	3 ml	5, 6, 11, 12, 14
(1-8)	2 M NaAc (pH 6.5)	1 ml	5, 6, 7, 8, 10, 12, 14
(1-9)	GTE buffer (pH 7.5) 0.05 M glucose◊ 0.025 M Tris (pH 7.5)☆ 0.01 M EDTA☆	2.4 ml	6, 10, 13
(1-10)	Lysis solution 0.2 N NaOH 1% (w/v) SDS☆	4.8 ml	6, 10, 13
(1-11)	5 M KAc buffer (pH 4.8) 29.4 g KAc 12 ml glacial acetic acid H$_2$O to 100 ml (Note: the pH may vary somewhat between 4.8–5.0)	4.2 ml	6, 10, 13
(1-12)	TE buffer (pH 7.4) 0.01 M Tris (pH 7.4)☆ 0.001 M EDTA☆	3.2 ml	6, 7, 8, 10, 14

☆Included in Chemicals Package (see page 163)   ⊕Included in Biological Package (see page 163)
◊Included in Media Package (see page 163)

Solution	Amounts Required	Periods Used
(1-13) 10× *Eco*RI buffer (pH 7.2) 0.5 M Tris (pH 7.2)☆ 0.1 M $MgCl_2$ 0.5 M NaCl 0.02 M dithiothreitol☆ Store at −20°C	65 μl	7, 12, 14
(1-14) 20× TAE electrophoresis buffer (pH 7.2) 0.8 M Tris (pH 7.2)☆ 0.3 M NaAc 0.02 M EDTA☆	400 ml	7, 8, 10, 11, 13, 17
(1-15) Sample solution 20% (w/v) ficoll (400,000 m.w.) or sucrose☆ 0.2 mg bromophenol blue/ml☆	0.25 ml	7, 8, 10, 11, 13
(1-16) Ethidium bromide☆ staining solution (class solution—used for the entire course) 0.5 mg ethidium bromide/500 ml H_2O	200 ml	7, 8, 10, 11, 13, 17
(1-17) 0.01 M ATP (store at −20°C)☆	26 μl	7, 14
(1-18) 10× ligation buffer (pH 7.2) 0.5 M Tris (pH 7.2)☆ 0.1 M $MgCl_2$ 0.01 M dithiothreitol☆ Store at −20°C	26 μl	7, 14
(1-19) 0.05M $CaCl_2$	38 ml	8, 15, 19
(1-20) [Omitted]		
(1-21) Tetracycline at 20 mg/ml in H_2O◇ Make up fresh before use	75 μl	9
(1-22) 10× Bam buffer (pH 8.0) 1.0 M NaCl 0.5 M Tris (pH 8.0)☆ 0.1 M $MgCl_2$ 0.2 M dithiothreitol☆ Store at −20°C	40 μl	10
(1-23) 0.1 M isopropylthiogalactoside☆ (IPTG) in H_2O Store at −20°C	0.11 ml	11, 15, 19
(1-24) 2% (w/v) X*gal* (5-chloro-4-bromo- 3-indoyl-β-D-galactoside) in dimethyl formamide☆ Store at 4°C	0.55 ml	11, 15, 19
(1-25) 0.9% (w/v) NaCl 0.9 g NaCl 100 ml H_2O	40 ml	11
(1-26) Ampicillin at 50 mg/ml in H_2O◇ Make up fresh before use	1 μl	11

☆Included in Chemicals Package (see page 163)   ◇Included in Media Package (see page 163)

	Solution	Amounts Required	Periods Used
(1-26)	Ampicillin at 50 mg/ml in H_2O◊ Make up fresh before use	1 μl	11
(1-27)	Chloramphenicol at 17 mg/ml in◊ ethanol	0.35 ml	12
(1-28)	0.09 M EDTA (pH 8.3)☆ (a) Use the tetra sodium salt. (b) Adjust the pH with acetic acid or NaOH.	30 μl	12, 14
(1-29)	Low-Tris buffer (pH 7.5) 0.01 M NaCl 0.01 M Tris (pH 7.5)☆ 0.001 M EDTA☆	0.17 ml	14, 15, 19
(1-30)	0.05 M $CaCl_2$	31 ml	15, 19
(1-31)	Loading buffer (pH 8.3) 0.02% (w/v) bromophenol blue☆ 0.2 M EDTA (pH 8.3) 20% (w/v) ficoll☆	72 μl	17
(1-32)	2% (w/v) SDS☆ (be careful not to inhale SDS—it induces a terrible coughing spell)	20 μl	17
(1-33)	10× *Hae*II buffer (pH 8.1) 0.5 M Tris (pH 8.1)☆ 0.5 M NaCl 0.1 M $MgCl_2$ 0.02 M dithiothreitol☆ Store at $-20°$C	4 μl	18
(1-34)	20× SSC (pH 7.2) Per liter: 175 g NaCl 88.2 g sodium citrate adjust pH with drops of 10 N NaOH	20 μl	18

2. CHEMICAL REAGENTS

	Reagent	Amounts Required	Periods Used
(2-1)	Chloroform	2.5 ml	3, 5, 6
(2-2)	Isopropanol	30 ml	6, 7, 8, 10, 12, 13, 14
(2-3)	Ethanol (95%)	2 ml	5, 6
(2-4)	Agarose☆	2.1 g	7, 8, 10, 11, 13, 17
(2-5)	Dimethylsulfoxide (DMSO)	9 μl	8
(2-6)	Mineral Oil	220 μl	17, 18

☆Included in Chemicals Package (see page 163)   ◊Included in Media Package (see page 163)

3. CULTURE MEDIA AND AGAR PLATES

(3-1)   RB = Rich broth medium (700 ml required; Periods 1, 2, 4, 5, 7, 9, 10, 11, 12, 14, 16, 18)

Per liter: 10 g tryptone◊
           5 g yeast extract◊
        10 g NaCl

(a) Autoclave.
(b) Add 10 ml 0.4% (w/v) thymine◊—autoclave separately.
(c) Add 5 ml 40% (w/v) glucose◊—autoclave separately.

(3-1/amp)	(3-1/cap)	(3-1/Hyu)	(3-1/spc)	(3-1/tet)
RB/amp	RB/cap	RB/Hyu	RB/spc	RB/tet

Drug-supplemented RB media contain one or more of the following antibiotics at the concentrations listed:

RB/amp = 50 μg ampicillin/ml RB
RB/cap = 20 μg chloramphenicol/ml RB
RB/Hyu = 750 μg hydroxyurea/ml RB (sometimes 2.0 mg/ml)
RB/spc = 30 μg spectinomycin/ml RB
RB/tet = 20 μg tetracycline/ml RB

(3-2)   RA = Rich agar   (80 plates required; Periods 1, 2, 3, 4, 8, 9, 11, 15, 19)
1000 ml RB (3-1)◊
    15 g agar◊

(a) Autoclave the Bactotryptone, yeast extract, NaCl and agar together.
(b) Add the thymine and glucose as described above.
(c) Pour about 25 ml of agar into each 100-mm plastic plate, except those for phage titers which need 30–40 ml.
(d) Flame the surface of the agar after pouring, to remove any air bubbles.

(3-2/amp)	(3-2/cap)	(3-2/Hyu)	(3-2/tet)
RA/amp	RA/cap	RA/Hyu	RA/tet

(a) Add drug supplements to the agar plates at Step (b) of (3-2) after the agar has cooled to 55°C.
(b) Drugs should be added at the same concentrations as given above for the drug-supplemented RB media.
(c) Plates containing antibiotics should be stored at 4°C and used within three days. Tetracycline plates should be stored in the dark due to the light sensitivity of the antibiotic.

(3-2/s)   Soft agar   (+ drug supplements, e.g., 3-2/amp-s)
(75 ml required; Periods 4, 11, 15, 19)

(a) Follow the procedure for (3-2) except use only 6.5 g agar/l rather than 15 g/l.

(3-3)   SB = Super broth medium   (50 ml required per team plus 400 ml for the entire class; Periods 2, 3)

Per liter: 33 g tryptone◊
        20 g yeast extract◊
       7.5 g NaCl
       3.5 ml 10 N NaOH
       2.0 g glucose◊
       20 mg thymidine◊

◊Included in Media Package (see page 163)

(3-4)   MB = Maltose broth medium   (125 ml required; Periods 3, 4)

    (a) Follow the procedure for (3-1) except substitute 5 ml 40% (w/v) maltose◊ for the glucose.

(3-5)   Z broth   **(9 ml required; Period 8)**

Per liter: 16 g nutrient broth◊
            10 g peptone◊
            2 g glucose◊

(3-6)   Minimal glucose plates   (one plate required, Period 10, for growth of JM107 by instructor)

Per liter: 100 ml 10× salts
            1 ml 1 M $MgSO_4$
          0.1 ml 1 M $CaCl_2$
          0.1 ml 1% vitamin B1◊
         10 ml 20% (w/v) glucose◊
         15 g agar

10× salts   6 g NaH_2PO_4
          3 g Na_2HPO_4
       0.5 g NaCl
         1 g NH_4Cl

    (a) Autoclave the 10× salts, $MgSO_4$, and $CaCl_2$ solutions separately.
    (b) Filter and sterilize the vitamin B1 and glucose solutions separately.
    (c) Autoclave 16 g agar in 889 ml H_2O.
    (d) Combine the solutions while the agar is hot, swirl to mix and pour plates. Flame the agar surfaces as described for (3-2).

4. BIOLOGICAL REAGENTS

	Biological Reagent	Amounts Required	Periods Used
(4-1)	E. coli JF335⊕	Stab or slant	1
(4-2)	E. coli JF413⊕	Stab or slant	1
(4-3)	E. coli JF417⊕	Stab or slant	1
(4-4)	E. coli JF427⊕	Stab or slant	1
(4-5)	E. coli JF428⊕	Stab or slant	5, 8
(4-6)	E. coli JF429⊕	Stab or slant	8
(4-7)	E. coli JM107⊕	Stab or slant	10, 11, 16, 19
(4-8)	Phage M13mp10⊕	50 μl, 10^{11}–10^{12}/ml	11
(4-8a)	Phage M13mp10 RF⊕	1 μg	12
(4-9)	Phage M13mp11⊕	50 μl, 10^{11}–10^{12}/ml	11
(4-9a)	Phage M13mp11 RF⊕	1 μg	12
(4-10)	Phage M13mp10-16⊕	0.1 μg	18
(4-11)	Plasmid pUC12⊕	2 μg	18
(4-12)	Endonuclease EcoRI⊕	40 units (4 μl)	7
(4-13)	T4 Ligase⊕	10 units (10 μl)	7, 14
(4-14)	Endonuclease BamHI⊕	120 units (12 μl)	10, 12
(4-15)	Endonuclease SstI⊕	30 units (30 μl)	12, 14

⊕ Included in Biological Package (see page 163)   ◊ Included in Media Package (see page 163)

	Biological Reagent	Amounts Required	Periods Used
(4-16)	Endonuclease *BglII*$^\oplus$	25 units (2.5µl)	14
(4-17)	Endonuclease *HaeII*$^\oplus$	5 units (2 µl)	18
(4-18)	Lysozyme$^\oplus$	0.8 mg	6, 10, 13
(4-19)	Phage λ DNA*$^\oplus$	—	—
(4-20)	Phage λd*nrd* DNA*$^\oplus$	—	—
(4-21)	Plasmid pBR325*$^\oplus$	—	—
(4-22)	Plasmid pBR*nrd*$^\oplus$	—	—

*These recombinant DNAs are made in class. They are available in case a particular procedure is unsuccessful or for laboratory courses that do not include the entire set of experiments and therefore need one or more of the intermediates. These DNAs are also useful as markers for gel electrophoresis and as biological standards against which DNAs prepared in class can be compared.

$\oplus$ Included in Biological Package (see page 163)

5. SUPPLIES

1. One 50-ml Erlenmeyer flask

2. Two 250-ml Erlenmeyer flasks

3. Two 1000-ml Erlenmeyer flasks

4. One 100-ml graduated cylinder

5. One 1000-ml graduated cylinder

6. Assorted beakers

7. Twenty-five 25-ml dilution tubes (10 caps are needed)

8. About 100 5-ml disposable tubes (50 caps are needed)

9. 1 ml, 5 ml, 10 ml pipettes

10. Pipette cans

11. Micropipettes (0–20 µl and 20–200 µl; Gilson "Pipetteman" micropipettes are available from Ranin Instruments, Woburn, MA 01801)

12. Three 15-ml Corex centrifuge tubes

13. Two 50-ml polypropylene centrifuge tubes

14. 100 1.5-ml plastic minifuge tubes

15. 150 0.5 ml plastic minifuge tubes

16. Disposable micropipette tips

17. Disposable petri dishes (100 × 15 mm)

18. Disposable Pasteur pipettes (9 inches long)

19. Disposable gloves

20. Safety goggles

21. $7\frac{1}{2} \times 5\frac{1}{2} \times 2$ inch plastic tray with lid (normally used for storing food; we use it for the class ethidium bromide staining solution)

22. Bunsen burner, ring stand, asbestos pad

23. Inoculating loop

24. Magnetic spin bars

25. Ice bucket

26. One 500-ml side-arm Erlenmeyer flask with tygon tubing (for aspirating alcohol; see Note 5, Period 7)

27. Polaroid film

28. Whatman #3 paper

29. Toothpicks

30. Two- and three-cycle semilog paper

6. EQUIPMENT

1. Two shaker baths or shaking incubation chambers (for 30°C, 37°C, and 42°C incubations)

2. Centrifuge (Sorvall RC5B or equivalent)

3. Culture tube rocker, roller, or shaker for incubation of 25-ml culture tubes at 37°C

4. 30°C, 37°C, and 42°C incubator for agar plates (not required simultaneously)

5. 4°C refrigerator/−20°C freezer

6. Electrophoresis apparatus
   (a) Homemade gel boxes; see Appendix B or equivalent commercial supplier
   (b) Power supply 150 V at 30 A (Heathkit or equivalent power supplies)

7. Magnetic stirrer with spin bars

8. Polaroid camera apparatus (see page 92) with transilluminator

9. Vacuum source or art supply (see Period 7, Notes 1 and 2)

7. SOURCE OF REAGENT PACKAGES

Bethesda Research Laboratories, Inc. (BRL) is the exclusive supplier of the three reagent packages referred to in the previous pages. The Biological Package, the Chemicals Package, and the Media Package are designed to be used by 10 pairs of students and contain all the components necessary for this course that are not usually found in a teaching laboratory. The packages contain a sufficient amount of each component to allow for repetitions of experiments. Each of the three packages can be purchased individually from BRL. For further information about the reagent packages, call BRL at (800) 638-4045. From Maryland or outside the United States, call (301) 840-8000.

List of Components in each Reagent Package

Biological Package	Chemicals Package	Media Package
DNase I	Agarose	Agar
E. coli strain JF335	ATP	Ampicillin
E. coli strain JF413	Bromophenol blue	Chloramphenicol
E. coli strain JF417	Dithiothreitol	Glucose
E. coli strain JF427	EDTA	Hydroxyurea
E. coli strain JF428	Ethidium bromide	Maltose
E. coli strain JF429	Ficoll	Nutrient broth
E. coli strain JM107	IPTG	Peptone
Endonuclease *Bam*HI	Lysozyme	Spectinomycin
Endonuclease *Bgl*II	MOPS	Tetracycline
Endonuclease *Eco*RI	Phenol	Thymine
Endonuclease *Hae*II	RbCl	Tryptone
Endonuclease *Sst*I	Sodium Dodecyl Sulfate (SDS)	Vitamin B1
Lysozyme	Sucrose	Yeast extract
Phage M13mp10	Tris	
Phage M13mp10 RF DNA	*Xgal*	
Phage M13mp11		
Phage M13mp11 RF DNA		
Phage M13mp10-16 ss DNA		
Plasmid pBR*nrd* DNA		
Plasmid pUC12 DNA		
RNase T1		
T4 DNA ligase		

8. SOURCES OF MATERIALS (AND USEFUL, *FREE*, CATALOGUES)

(1) General Chemicals—(the companies we use most commonly)

    (a) Sigma Chemical Company   (they have the most extensive
        PO Box 14508   catalogue)
        St. Louis, MO   63178

    (b) Mallinkrodt, Inc.
        675 Brown Road
        PO Box 5840
        St. Louis, MO   63134

    (c) Calbio Chem-Behring
        PO Box 12087
        San Diego, CA   92112

(d) BioRad Laboratories
   2200 Wright Avenue
   Richmond, CA 94804

(2) Culture Media

   (a) GIBCO Laboratories, Inc.
       Madison, WI 53711

   (b) Difco Laboratories
       Detroit, MI 48201

   (c) Sigma Chemical Company

(3) Biological Reagents—(there is an evergrowing number of suppliers of reagents used in recombinant DNA work)

   (a) Bethesda Research Laboratories
       PO Box 6009
       Gaithersburg, MD 20877

   (b) New England Biolabs, Inc.
       32 Tozer Road
       Beverly, MA 01915

   (c) Boehringer Mannheim Biochemicals
       7941 Castleway Drive
       PO Box 50816
       Indianapolis, IN 46250

   (d) International Biotechnologies, Inc.
       PO Box 1565
       New Haven, CT 06506

(4) Supplies and Equipment

   (a) American Scientific Products
       1210 Waukegan Road
       McGaw Park, IL 60085
       (Plus many branch offices; they carry Difco products)

   (b) Sargent-Welch Scientific Company
       7300 N. Linder Avenue
       PO Box 1026
       Skokie, IL 60077
       (Plus many branch offices)

   (c) A. H. Thomas Company
       PO Box 779
       Philadelphia, PA 19105

   (d) Fischer Scientific Company
       1600 W. Glenlake Avenue
       Itasca, IL 60143
       (Plus many other offices)

(e) Sorvall Centrifuges
Dupont Company
Instrument Products
Biomedical Division
Newtown, CT   60470

Appendix B

Homemade Laboratory Equipment

CONSTRUCTION OF A CHEAP, SIMPLE BUT EFFECTIVE PROPIPETTE (Figure B-1)

Two types of propipettes will be useful for this course and can be simply constructed. One that is useful for disposable pipettes (Pasteur) is constructed from a 3 cc or larger disposable syringe, a Pipetteman tip (0–20 μl —yellow tip) and a 1-inch piece of latex tubing (1/8-inch inner diameter and 1/16-inch thick wall). Cut off 3/8 inch from the large end of the pipette tip and place the shorter pipette tip on the syringe as firmly as possible. Place the pipette tip through the latex tubing and force the tubing as far up the tip as possible. When this tip is firmly inserted into a disposable pipette, the friction will hold the pipette. Liquid is drawn up and expelled by using the thumb nail to force the plunger up, and the flesh side of the thumb to force the plunger down. This device is very useful in allowing one to use disposable pipettes to carry out spectrophotometric determinations of cell growth, using approximate 1-ml volumes as determined by the syringe.

For use as a propipette for 1-ml, 5-ml, and 10-ml pipettes, a similar device can be used. Use a 12 cc disposable syringe and a Pipetteman tip (0–20 μl—yellow tip) with 3/8 inch of the top cut off. No latex tubing is needed since the inner diameter of the top of a pipette is small enough that the pipette tip makes a seal. Due to the weight of the larger pipette filled with liquid, two hands may be required to operate this device. Use this device to draw liquid above the desired level; remove the device and use finger control to release the desired amount of liquid.

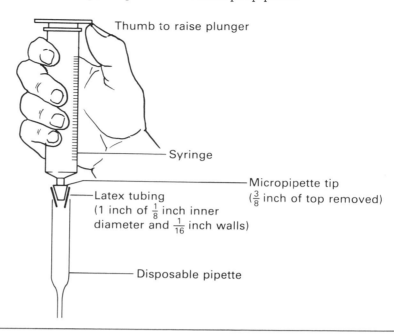

Figure B-1 A cheap, simple and efficient propipette.

Thumb to raise plunger

Syringe

Micropipette tip
($\frac{3}{8}$ inch of top removed)

Latex tubing
(1 inch of $\frac{1}{8}$ inch inner
diameter and $\frac{1}{16}$ inch walls)

Disposable pipette

CONSTRUCTION OF GEL ELECTROPHORESIS EQUIPMENT

Glass Plates

Kodak lantern slide cover glasses ($3\frac{1}{4} \times 4$ inches or 8.3×10.2 cm)

Electrophoresis Buffer Reservoirs—(To Hold 500 ml Buffer)

1. Rubbermaid brand drawer organizers, or other plastic boxes $9 \times 3 \times 2$ inches. Sold in the kitchen section of general stores for about a dollar each.

2. Three lucite blocks

   Two blocks $1/4 \times 1/4 \times 1/2$ inch with a small hole
   One block $3/4 \times 3/4 \times 1/2$ inch with a hole for a banana plug

3. One banana plug (screw type with nut)

4. 7–8 inches platinum wire (24 guage) (90% the cost of the reservoir)

5. Assemble as shown (use dichloromethylene to glue lucite to plastic)

Figure B-2 Construction of a reservoir for agarose minigels.

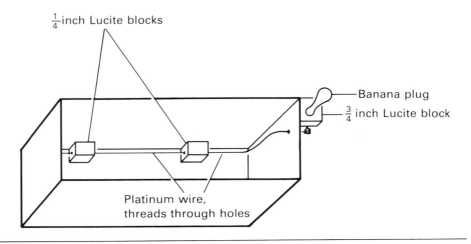

C. Well-Former

Figure B 3 Well-former for agarose minigels.

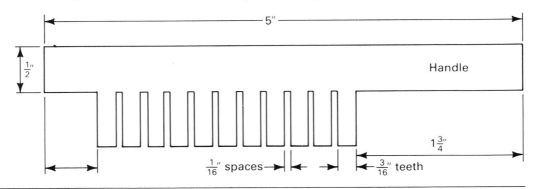

1. Make from 1/16-inch lucite

Well-former holder

1. Twelve standard (1 × 3 inches) microscope slides
2. Tape slides together as shown (top view)

Figure B-4 Well-former holder.

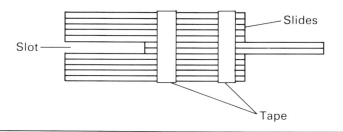

3. Put handle of well-former into slot of holder and put on glass slide

Figure B-5 Setup for pouring agarose minigels.

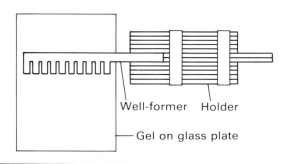

Appendix C
Relative Centrifugal Force Chart

All centrifugation conditions listed in this manual are given in terms of force, in *g* units (or gravities), and time. We also list the speeds for centrifuging in Sorvall centrifuges with specific rotors. The chart in this appendix shows the relative forces at various speeds using rotors that hold centrifuge tubes at various distances (*r*, measured in *inches*) from the center of rotation. Regardless of centrifuge type, the above values are correct. Consequently, if a centrifuge and/or rotor other than that specified in the protocols is used, the experimenter can measure the *r* value for the rotor and determine the best speed for centrifugation.

RELATIVE CENTRIFUGAL FORCE CHART

$$\text{R.C.F.} = \frac{4(3.1416)^2 r n^2}{32.2} \times \text{gravity} \qquad \text{where } r = \text{radius (in feet)}$$

$$n = \text{revolutions per second} = \frac{\text{rpm}}{60}$$

SORVALL SS-34

Speed (rpm)	1	2.76	3	4.25	4.34	4.52	4.65	4.85	5.75	6	7.25	9.83
500	7	19	21	30	31	32	33	34	40	43	51	69
1,000	28	77	85	121	122	126	132	138	164	170	206	280
1,500	64	177	192	270	278	290	295	310	365	383	463	630
2,000	114	315	341	480	495	515	527	550	650	681	822	1120
2,500	177	490	532	755	770	800	825	860	1020	1060	1290	1740
3,000	255	700	766	1085	1110	1150	1185	1240	1465	1530	1850	2510
3,500	348	960	1040	1475	1510	1580	1600	1680	2000	2090	2520	3420
4,000	454	1250	1360	1935	1980	2050	2110	2200	2520	2720	3290	4470
4,500	575	1590	1720	2445	2500	2600	2675	2790	3300	3450	4170	5650
5,000	710	1960	2130	3020	3090	3200	3300	3440	4080	4260	5140	6975
5,500	860	2370	2580	3640	3730	3880	3980	4150	4920	5150	6220	
6,000	1020	2820	3070	4340	4430	4600	4740	4950	5860	6130	7410	
6,500	1200	3310	3600	5090	5210	5420	5560	5800	6870	7190	8690	
7,000	1390	3840	4170	5900	6040	6280	6450	6740	7970	8340	10100	
7,500	1600	4420	4790	6780	6950	7230	7400	7740	9150	9580	11600	
8,000	1820	5020	5450	7710	7900	8230	8440	8800	10400	10900	13200	
8,500	2050	5660	6150	8700	8900	9270	9500	9900	11700	12300	14900	
9,000	2300	6350	6900	9750	9990	10400	10660	11100	13200	13800	16700	
9,500	2560	7070	7680	10800	11100	11570	11800	12300	14600	15400	18600	
10,000	2840	7850	8520	12100	12350	12840	13200	13800	16300	17000	20600	
10,500	3130	8650	9390	13300	13600	14100	14500	15200	18000	18800	22700	
11,000	3430	9500	10300	14500	14900	15500	15900	16500	19600	20600	24900	
11,500	3750	10300	11300	15900	16300	16900	17400	18100	21500	22500	27200	
12,000	4090	11300	12300	17300	17750	18500	18900	19700	23300	24500	29600	
12,500	4430	12200	13300	18800	19250	20000	20300	21400	25400	26600	32100	
13,000	4800	13200	14400	20200	20850	21700	22100	23000	27300	28800	34800	
13,500	5170	14300	15500	22000	22450	23400	24000	25000	29700	31000	37500	
14,000	5560	15400	16700	23500	24150	25100	25700	26800	31700	33400	40300	
14,500	5970	16500	17900	25300	25900	27000	27700	28800	34200	35800	43300	
15,000	6390	17600	19200	27000	27750	28900	29500	30800	36400	38300	46300	
15,500	6820	18800	20500	29000	29600	30800	31700	33000	39200	40900	49400	
16,000	7270	20100	21800	30900	31550	32900	33800	35200	41700	43600	52700	
16,500	7730	21400	23200	32800	33550	35000	35800	37400	44400	46400	56000	
17,000	8200	22700	24600	34800	35600	37100	38000	39700	47000	49200	59500	
17,500	8690	24000	26100	37000	37750	39300	40400	42200	50000	52100	63000	
18,000	9190	25400	27600	39100	39900	41500	42700	44500	52800	55200	66700	
18,500	9710	26800	29100	41300	42200	43900	45100	47000	55800	58300	70400	
19,000	10200	28200	30700	43500	44300	46100	47500	49600	58700	61500	74300	
19,500	10800	30800	32400	45900	46900	48800	50200	52300	62000	64700	78200	
20,000	11400	31500	34100	48200	49500	51500	52600	55000	65000	68100	82300	

Appendix D
Notes on DNA Analysis by Gel Electrophoresis

In this laboratory, we shall analyze DNA molecules and fragments on slab gels of agarose. We shall use minigels that consist of 25 ml of an agarose solution, usually 0.8–1.0% (w/v) agarose. The hot agarose is poured onto a $3\frac{1}{4} \times 4$ inches (83×102 mm) glass lantern slide and allowed to solidify with a comb or well-former in place. The gel, containing an appropriate buffer (see below) that is identical to the electrophoresis buffer, acts as a conduction bridge between the two reservoirs. DNA samples are loaded into wells situated close to the cathode ($-$ pole) and pulled, by an applied electric field, through the gel toward the anode ($+$ pole).

The rate of movement of the DNA molecules through agarose, or polyacrylamide, gels is a function of (a) the concentration and type of the gel matrix—the pore size of the agarose or acrylamide will sieve the DNA molecules; accordingly, a 1% agarose gel, which has a relatively loose matrix, is used for the separation of molecules 1–60 kilobase pairs (0.6–40 $\times 10^6$ mol. wt.) whereas an 8% polyacrylamide gel, which has a tighter matrix, is useful for DNAs ranging in size from 100–3000 base pairs (0.03–2.0 $\times 10^6$ mol. wt.). (b) The conformation of the nucleic acids being separated; double-stranded DNA normally exists in the three following conformations with decreasing relative mobilities: covalently closed superhelices, linear, kinky coils and open circles formed by closing linear DNAs or nicking one strand of a superhelical form. In addition, double-stranded DNA can be denatured into single strands that will form random coils; these structures migrate at different rates from the three double-stranded forms. (c) Strength of the electric field, that is, the voltage. (d) Temperature at which the gels are run; (e) pH of the gel buffer; at various pH's the different bases are differentially charged, and thus nucleic acids of equal length but different base composition can migrate at different rates. DNA fragments with similar conformations migrate according to the relationship log (mol. wt.) α distance migrated in the center of the gel. Consequently, the size of a DNA molecule can be determined from its migration in a gel relative to those of marker DNAs of known molecular weights.

Figure D-1 shows a 0.8% agarose gel in TAE buffer (pH 7.5) through

Figure D-1 DNA fragments separated in an agarose minigel. Various DNA samples used in the course were electrophoresed through a 0.8% (w/v) agarose gel and stained with ethidium bromide. The DNA fragments are identified in the text.

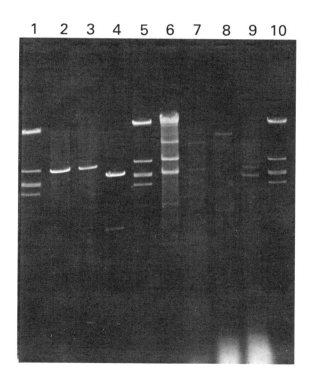

which several DNAs used in this laboratory were electrophoresed. The DNAs and their fragment sizes (in kilobase pairs) are:

Lanes 1, 5, 10: λ helper × *Eco*RI → 21.76, 7.75, 5.88, 5.54, 4.85, 3.43 (invisible); see page 22 for the restriction map.

Lane 2: M13mp10 RF × *Sst*I and *Bam*HI → 7.2; see page 22 for the map.

Lane 3: M13mp11 RF × *Sst*I and *Bam*HI → 7.2; see page 41 for the map.

Lane 4: pBR325 × *Eco*RI → 6.0 (plus uncut supercoil with a mobility of 4.3 due to its compact conformation), see page 26 for the map.

Lane 6: λα*nrd* × *Eco*RI → 18.5, 10.94, 7.55, 5.88, 3.43 (barely visible) see page 22 for the restriction map.

Lane 7: pBR*nrd* × *Eco*RI → 10.9, 6.0. pBR*nrd* is identified on page 82.

Lane 8: pBR*nrd* × *Sst*I and *Bgl*II → 13.5, 3.1, 0.4 (invisible).

Lane 9: pBR*nrd* × *Bam*HI → 6.2, 5.4, 5.4 (2 bands superimposed).

The above identified fragments have been graphed on two-cycle semi-log paper in Figure D-2. This graph indicates that there are two regions where the fragments migrate as a function of the log (size—in base pairs or molecular weight). Clearly the λ × *Eco*RI fragments are suitable markers for the DNAs used in this class.

A variety of electrophoretic systems employing different buffers is used for different purposes. We use a Tris-acetate buffer which is among the most commonly used for DNA sizing. In addition, there are Tris-borate buffers for methyl mercuric hydroxide denatured nucleic acids (Bailey and Davidson, 1976), and phosphate buffers for glyoxal denatured nucleic acids (McMaster and Carmichael, 1977). The merits of the various buffer systems is discussed in Davis, Botstein and Roth (1980).

Figure D-2 Semilog plot of the DNA fragments shown in Figure D-1 showing the electrophoretic mobility as a function of fragment size.

● : λ helper × *Eco*RI.
○ : λd*nrd* × *Eco*RI.
✕ : pBR325 × *Eco*RI.
* : M13mp10 RF (and M13mp11 RF) × *Sst*I and *Bam*HI.
△ : pBR*nrd* × *Eco*RI.
▲ : pBR*nrd* × *Sst*I and *Bgl*II.
□ : pBR*nrd* × *Bam*HI.

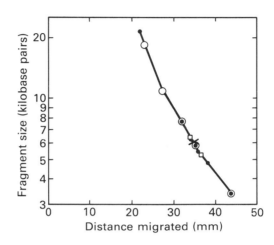

Appendix E
Notes on DNA Ligation

Seven parameters affect the rate of joining of DNA fragments: the DNA concentration, the concentration of compatible ends, the size of the fragments, the presence or absence of complementary, single-stranded ends (some nucleases form blunt ends and in these cases special techniques are needed for joining), the G + C content and length of the complementary single strands, the temperature of the reaction, and the ionic composition of the solution. In this section, we will consider only the case of joining of molecules having single-stranded termini (*sticky-end ligation*).

The strength of the hydrogen bonding between compatible sticky ends depends on their length and G + C content. The melting temperature (T_m) of AT pairs is much lower than that of GC pairs. In the cloning of *Eco*RI fragments of λ DNA into plasmid pBR325 the sequence of each single-stranded terminus is AATT; the value of T_m for the double-stranded joint formed by annealing these ends is 4–5°C. After allowing hydrogen bonds to form, the joint is covalently sealed with T4 DNA ligase, an enzyme whose activity is maximal at 37°C, a temperature at which the AATT double-stranded region would be disrupted. Fortunately, the enzyme will function at lower temperatures, albeit slowly. At 12°C, the double-stranded AATT joint is very unstable but not totally so, for it is being made and broken continually. Thus, if sufficient time is allowed at this temperature, the T4 DNA ligase succeeds in sealing the joints in the double-stranded segments that exist at a particular instant. Therefore, this temperature is usually selected as a convenient compromise between rate and efficiency.

The configuration of single-stranded DNA is affected by ionic conditions. At high ionic strength single strands collapse and form instrastrand hydrogen bonds, which prevent intermolecular pairing. At low ionic strength single strands are extended, but electrostatic repulsion between the highly charged phosphate groups prevents the strands from coming into contact. Furthermore, owing to more subtle effects on the shape of double-stranded DNA, intramolecular joining is facilitated by high ionic strength. In some experiments, both intermolecular and intramolecular joining may be needed, the former to join different linear fragments and the latter for circularization of DNA prior to transformation. In the procedures used in this manual, moderately low concentrations of monovalent and divalent cations—50 mM and 10 mM, respectively—are used. At these concentrations single strands are somewhat extended yet the charge is neutralized sufficiently to allow approach of the strands.

A simple mathematical treatment, which follows, enables the outcome of the annealing reaction to be analyzed as a function of several parameters. In this treatment, the total concentration of *all* DNA termini is denoted i and the effective concentration of the two ends on the *same* molecule (one end in the neighborhood of the other) by j. Thus, j is inversely proportional to the length of the particular DNA fragment, and its value relative to i will determine the extent to which circularization (intramolecular joining) occurs compared to coupling to a second fragment (intermolecular joining). Since the shape of a DNA fragment is influenced by the ionic strength of the environment, j is sensitive to ion concentration during the annealing reaction. Using λ DNA as a standard,

$$j_x = j_\lambda \left(\frac{M_\lambda}{M_x}\right)^{3/2}$$

(1)

for any other fragment x having molecular weight M_x. For λ, $j_\lambda = 3.4 \times 10^{11}$ ends/ml and $M_\lambda = 32 \times 10^6$. The 3/2 power of the ratio of the M values is the result of the flexibility of DNA fragments—the separation of the ends of a fragment is not simply proportional to the length of the fragment. The concentration of sticky ends i_x in the annealing reaction is twice the concentration of fragments (each linear fragment has two ends) or, in units of ends/ml,

$$i_x = 2N_A M \times 10^{-3} \text{ ends/ml}$$

(2)

where N_A is the Avogadro number (6.02×10^{23} molecules/mole), M is the molarity of DNA fragments, and the factor 10^3 converts liters to milliliters.

For *equal* rates of intermolecular and intramolecular joining, one end of a fragment should react with its other end or with that of another fragment at equal rates; thus, $i = j$, or $j/i = 1$. Combining equations (1) and (2), we get

$$\frac{j_x}{i_x} = \frac{5.1 \times 10^4}{[\text{DNA}_x](M_x)^{1/2}}$$

(3)

where [DNA$_x$] is the concentration of DNA expressed in micrograms per milliliter.

This equation applies to reactions in which all fragments have the same size. If the DNA concentration increases, linear concatemers will form more rapidly than will circularized molecules and, conversely, when the DNA concentration is lowered ($j > i$), intramolecular ligation is favored. If the joining reaction proceeds to completion, then all molecules should eventually circularize. When $i > j$, the circles should be larger than when $j > i$. The derivation is much more complex if the reaction mixture contains fragments of many sizes. Generally, two procedures are utilized to get a *rough* estimate of the outcome of an annealing reaction—that is, whether intermolecular or intramolecular reactions will occur. These procedures are as follows: (1) the sizes of the fragments are averaged and the j/i ratio is calculated from equation (3), or a particular fragment size is chosen. (2) A graph of DNA concentration versus molecular weight is constructed and compared to the set of theoretical curves depicted in Figure E-1 [from Dugaiczyk et al. (1975), with permission]. Then, the approximate value of j/i is determined by interpolation or extrapolation from the theoretical curves. The results obtained by either of these procedures is, of course, an approximation because many fragments interact with themselves and with others of different size, thereby distorting the pure reactions on which the equations above are based. However, in practice, the results are useful.

Figure E-1 Approximate j/i values for DNA fragments at various concentrations. The graph shows the j/i value for DNAs with molecular weights between 0–6 × 10^6, 0–9000 base pairs, at normal ligation concentrations. j/i values between 0.3 and 1.5 are used for efficient ligation.

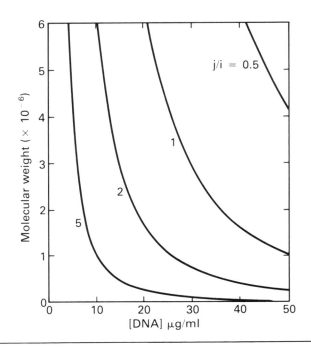

Appendix F
Notes on the Use of Restriction Enzymes

In this appendix, various guidelines for handling restriction enzymes are presented. In general, the following should be kept in mind. (a) Restriction enzymes are expensive; follow the rules given below (Factors which Affect Restriction Enzyme Activity). (b) Restriction enzymes are unstable; store them at -20° C, keep stock solutions on ice when they are out of storage. (c) Restriction enzymes are specific; stock solutions must *never* be contaminated, so use a fresh Pipetteman tip whenever you remove enzyme from the stock. (d) Consult product sheets for the correct handling of all restriction enzymes.

FACTORS WHICH AFFECT RESTRICTION ENZYME ACTIVITY

Summarized below are general suggestions on how to achieve optimal activity from these enzymes.

1. Follow the instructions on the Product Profile Sheets for storage and reaction conditions.

2. Use distilled water and enzyme grade reagents for all restriction buffers.

3. Several factors may influence the amount of enzyme required to digest different DNAs.

(a) The base composition adjacent to the enzyme recognition sequence can affect the digestion rate as much as 25 times.

(b) The density of cleavage sites in a particular DNA. For example, 1 μg of λ DNA has 0.06 pmoles of *Kpn*I cleavage sites whereas 1 μg of adenovirus-2 DNA has 0.36 pmoles of sites.

(c) Superhelical DNA may require more enzyme to obtain complete digestion than an equivalent population of linear molecules.

4. Agents which decrease the dielectric constant of the incubation medium can (reversibly) alter the cleavage specificity of restriction enzymes, particularly *Bam*HI, *Eco*RI, *Kpn*I, *Sau*3A, *Sal*I, *Hha*I, *Pst*I, *Sst*I and *Dde*I. Examples of reagents that produce this result are ethanol, glycerol, and DMSO. Glycerol (50%) often is used as the antifreeze component in restriction enzyme storage buffer; the final glycerol concentration in cleavage reactions should not be greater than 4%.

5. One should not assume that increasing digestion time (beyond one hour) will compensate for reducing the amount of enzyme added to a reaction.

6. Frequently, there is an increased enzyme activity when the DNA concentration is greater than 1 μg/50 μl. Thus, depending on your requirements, it may be worthwhile to determine how many micrograms of your DNA can be cleaved by one unit of enzyme.

7. Restriction enzymes and DNA fragments tend to stick to glass. Siliconized glassware or plastic should be used in all restriction enzyme reactions. Additionally, it is recommended that bovine serum albumin (BSA) or gelatin be added to solutions, where applicable, in order to stabilize the enzymes and avoid nonspecific absorption of macromolecules to the glassware.

8. Repeated freezing and thawing of DNAs should be avoided.

9. DNAs should not alter significantly the pH and/or the salt environment upon addition to reaction mixtures.

10. DNAs should be free of heavy metals which could inhibit the enzymes.

The following is a current list of identified restriction enzymes compiled by R. J. Roberts of the Cold Spring Harbor Laboratory and is reprinted with permission of Dr. Richard J. Roberts and *Nucleic Acids Research*, Vol. 10 No. 5, 1982. Table notes (a)–(h) are given on page 188.

Microorganism	Source	Enzyme[a]	Sequence[b]	λ	Ad2	SV40	φX	pBR
Acetobacter aceti sub. *liquefaciens*	IFO 12388	*Aac*I (*Bam*HI)	GGATCC	5	3	1	0	1
Acetobacter aceti sub. *liquefaciens*	M. Van Montagu	*Aae*I (*Bam*HI)	GGATCC	5	3	1	0	1
Acetobacter aceti sub. *orleanensis*	NCIB 8622	*Aor*I (*Eco*RII)	CC↓(^A_T)GG	>35	>35	16	2	6
Acetobacter pasteurianus sub. *pasteurianus*	NCIB 7215	*Apa*I	GGGCC↓C	1	>10	1	0	0
Achromobacter immobilis	ATCC 15934	*Aim*I	?	?	?	?	?	?
Acinetobacter calcoaceticus[h]	R.J. Roberts	*Acc*I	GT↓(^A_C)(^G_T)AC	7	8	1	2	2
		*Acc*II (*Fnu*DII)	CGCG	>50	>50	0	14	23
		*Acc*III	?	>10	>6	?	?	?
Agmenellum quadruplicatum	W.F. Doolittle	*Aqu*I (*Ava*I)	CPyCGPuG	8	15	0	1	1
Agrobacterium tumefaciens	ATCC 15955	*Atu*AI	?	>30	>30	?	?	?
Agrobacterium tumefaciens B6806	E. Nester	*Atu*BI (*Eco*RII)	CC(^A_T)GG	>35[d]	>35	16	2	6
Agrobacterium tumefaciens IIBV7	G. Roizes	*Atu*BVI	?	>14	?	1	0	?
Agrobacterium tumefaciens ID135	C. Kado	*Atu*II (*Eco*RII)	CC(^A_T)GG	>35[d]	>35	16	2	6
Agrobacterium tumefaciens C58	E. Nester	*Atu*CI (*Bcl*I)	TGATCA	7[d]	5	1	0	0
Alcaligenes species	N. Brown	*Asp*AI (*Bst*EII)	G↓GTNACC	11	8	0	0	0
Anabaena catanula	CCAP 1403/1	*Aca*I	?	?	?	?	?	?
Anabaena cylindrica	CCAP 1403/2a	*Acy*I	GPu↓CGPyC	>14	>14	0	7	6
Anabaena flos-aquae	A.E. Walsby	*Afl*I (*Ava*II)	G↓G(^A_T)CC	>17	>30	6	1	8
		*Afl*II	C↓TTAAG	3	?	0	2	0
		*Afl*III	?	?	?	?	?	?
Anabaena oscillarioides	CCAP 1403/11	*Aos*I (*Mst*I)	TGC↓GCA	>10	>15	0	1	4
		*Aos*II (*Acy*I)	GPu↓CGPyC	>14	>14	0	7	6
Anabaena strain Waterbury	ATCC 29208	*Ast*WI (*Acy*I)	GPu↓CGPyC	>14	>14	0	7	6
Anabaena subcylindrica	CCAP 1403/4b	*Asu*I	G↓GNCC	>30	>30	11	2	15
		*Asu*II	TT↓CGAA	7	1	0	0	0
		*Asu*III (*Acy*I)	GPu↓CGPyC	>14	>14	0	7	6
Anabaena variabilis	ATCC 27892	*Ava*I	C↓PyCGPuG	8	15	0	1	1
		*Ava*II	G↓G(^A_T)CC	>17	>30	6	1	8
Anabaena variabilis[uw]	E.C. Rosenvold	*Avr*I (*Ava*I)	CPyCGPuG	8	15	0	1	1
		*Avr*II	CCTAGG	2	2	2	0	0
Aphanothece halophytica	ATCC 29534	*Aha*I (*Cau*II)	CC(^C_G)GG	>30	>30	0	1	10
		*Aha*II	?	?	?	?	?	?
		*Aha*III	TTT↓AAA	13	>16	12	2	3
Arthrobacter luteus	ATCC 21606	*Alu*I	AG↓CT	>50	>50	35	24	16
Arthrobacter pyridinolis	R. DiLauro	*Apy*I (*Eco*RII)	CC↓(^A_T)GG	>35[d]	>35	16	2	6
Bacillus acidocaldarius	ATCC 27009	*Bac*I (*Sac*II)	CCGCGG	4	>25	0	1	0
Bacillus amyloliquefaciens F	ATCC 23350	*Bam*FI (*Bam*HI)	GGATCC	5	3	1	0	1
Bacillus amyloliquefaciens H	F.E. Young	*Bam*HI	G↓GATC̊C	5	3	1	0	1
Bacillus amyloliquefaciens K	T. Kaneko	*Bam*KI (*Bam*HI)	GGATCC	5	3	1	0	1
Bacillus amyloliquefaciens N	T. Ando	*Bam*NI (*Bam*HI)	GGATCC	5	3	1	0	1
		*Bam*N$_x$ (*Ava*II)	G↓G(^A_T)CC	>17	>30	6	1	8
Bacillus brevis S	A.P. Zarubina	*Bbv*SI	GC̊(^A_T)GC	specific methylase				
Bacillus brevis	ATCC 9999	*Bbv*I	GCAGC (8/12)	>30	>30	23	14	21
Bacillus caldolyticus	A. Atkinson	*Bcl*I	T↓GATCA	7[d]	5	1	0	0
Bacillus centrosporus	A.A. Janulaitis	*Bcn*I (*Nci*I)	CC↓GGG	>15	>15	0	1	10
Bacillus cereus	ATCC 14579	*Bce* 14579	?	>10	?	?	?	?
Bacillus cereus	IAM 1229	*Bce* 1229	?	>10	?	?	?	?

Microorganism	Source	Enzyme[a]	Sequence[b]	Number of cleavage sites[c]				
				λ	Ad2	SV40	φX	pBR
Bacillus cereus	T. Ando	*Bce* 170 (*Pst*I)	CTGCAG	18	25	2	1	1
Bacillus cereus Rf sm st	T. Ando	*Bce*R (*Fnu*DII)	CGCG	>50	>50	0	14	23
Bacillus globigii	G.A. Wilson	*Bgl*I	GCCNNNN↓NGGC	22	12	1	0	3
		*Bgl*II	A↓GATCT	6	12	0	0	0
Bacillus megaterium 899	B899	*Bme* 899	?	>5	?	?	?	?
Bacillus megaterium B205-3	T. Kaneko	*Bme* 205	?	>10	?	?	?	?
Bacillus megaterium	J. Upcroft	*Bme*I	?	>10	>20	4	?	?
Bacillus pumilus AHU1387A	T. Ando	*Bpu*I	?	6	>30	2	?	?
Bacillus sphaericus	IAM 1286	*Bsp* 1286	?	?	?	?	?	?
Bacillus sphaericus R	P. Venetianer	*Bsp*RI (*Hae*III)	GG↓CC	>50	>50	19	11	22
Bacillus stearothermophilus C1	N. Welker	*Bst*CI (*Hae*III)	GGCC	>50	>50	19	11	22
Bacillus stearothermophilus C11	N. Welker	*Bss*CI (*Hae*III)	GGCC	>50	>50	19	11	22
Bacillus stearothermophilus G3	N. Welker	*Bst*GI (*Bcl*I)	TGATCA	7[d]	5	1	0	0
		*Bst*GII (*Eco*RII)	CC(^A_T)GG	>35[d]	>35	16	2	6
Bacillus stearothermophilus G6	N. Welker	*Bss*GI (*Bst*XI)	?	10	9	0	3	0
		*Bss*GII (*Mbo*I)	GATC	>50[d]	>50	8	0	22
Bacillus stearothermophilus H1	N. Welker	*Bst*HI (*Xho*I)	CTCGAG	1	6	0	1	0
Bacillus stearothermophilus H3	N. Welker	*Bss*HI (*Xho*I)	CTCGAG	1	6	0	1	0
		*Bss*HII (*Bse*PI)	?	6	>18	0	2	0
Bacillus stearothermophilus H4	N. Welker	*Bsr*HI (*Bse*PI)	?	6	>18	0	2	0
Bacillus stearothermophilus P1	N. Welker	*Bss*PI	?	>8	?	?	?	?
Bacillus stearothermophilus P5	N. Welker	*Bsr*PI	?	11	>20	?	0	0
		*Bsr*PII	?	>50	?	?	?	?
Bacillus stearothermophilus P6	N. Welker	*Bse*PI	?	6	>18	0	2	0
Bacillus stearothermophilus P8	N. Welker	*Bsa*PI (*Mbo*I)	GATC	>50	>50	8	0	22
Bacillus stearothermophilus P9	N. Welker	*Bso*PI (*Bse*PI)	?	11	>20	?	0	0
Bacillus stearothermophilus T12	N. Welker	*Bst*TI (*Bst*XI)	?	10	9	0	3	0
Bacillus stearothermophilus X1	N. Welker	*Bst*XI	?	10	9	0	3	0
		*Bst*XII (*Mbo*I)	GATC	>50[d]	>50	8	0	22
Bacillus stearothermophilus 1503-4R	N. Welker	*Bst*I (*Bam*HI)	G↓GATCC	5	3	1	0	1
Bacillus stearothermophilus 240	A. Atkinson	*Bst*AI	?	?	?	?	?	?
Bacillus stearothermophilus ET	N. Welker	*Bst*EI	?	?	?	?	?	?
		*Bst*EII	G↓GTNACC	11	8	0	0	0
		*Bst*EIII (*Mbo*I)	GATC	>50[d]	>50	>8	0	22
Bacillus stearothermophilus	ATCC 12980	*Bst*PI (*Bst*EII)	G↓GTNACC	11	8	0	0	0
Bacillus stearothermophilus	D. Comb	*Bst*NI (*Eco*RII)	CC↓(^A_T)GG	>35[d]	>35	16	2	6
Bacillus stearothermophilus 822	G. Oshima	*Bse*I (*Hae*III)	GGCC	>50	>50	19	11	22
		*Bse*II (*Hpa*I)	GTTAAC	13	6	4	3	0
Bacillus subtilis strain R	T. Trautner	*Bsu*RI (*Hae*III)	GG↓C̊C	>50	>50	19	11	22
Bacillus subtilis Marburg 168	T. Ando	*Bsu*M	?	>10	?	?	?	?

Microorganism	Source	Enzyme[a]	Sequence[b]	Number of cleavage sites[c]				
				λ	Ad2	SV40	φX	pBR
Bacillus subtilis	ATCC 6633	*Bsu* 6663	?	>20	?	?	?	?
Bacillus subtilis	IAM 1076	*Bsu* 1076 (*Hae*III)	GGCC	>50	>50	19	11	22
Bacillus subtilis	IAM 1114	*Bsu* 1114 (*Hae*III)	GGCC	>50	>50	19	11	22
Bacillus subtilis	IAM 1247	*Bsu* 1247 (*Pst*I)	CTGCAG	18	25	2	1	1
Bacillus subtilis	ATCC 14593	*Bsu* 1145	?	>20	?	?	?	?
Bacillus subtilis	IAM 1192	*Bsu* 1192	?	>10	?	?	?	?
Bacillus subtilis	IAM 1193	*Bsu* 1193	?	>30	?	?	?	?
Bacillus subtilis	IAM 1231	*Bsu* 1231	?	>20	?	?	?	?
Bacillus subtilis	IAM 1259	*Bsu* 1259	?	>8	?	?	?	?
Bifidobacterium breve	H. Takahashi	*Bde*I (*Nar*I)	GGCGC↓C	>2	18	0	2	4
Bordetella bronchiseptica	ATCC 19395	*Bbr*I (*Hind*III)	AAGCTT	6	11	6	0	1
Bordetella pertussis	P. Novotny	*Bpe*I (*Hind*III)	AAGCTT	6	11	6	0	1
Brevibacterium albidum	ATCC 15831	*Bal*I	TGG↓C*CA	15	17	0	0	1
Brevibacterium luteum	ATCC 15830	*Blu*I (*Xho*I)	C↓TCGAG	1	6	0	1	0
		*Blu*II (*Hae*III)	GGCC	>50	>50	19	11	22
Calothrix scopulorum	CCAP 1410/5	*Csc*I (*Sac*II)	CCGC↓GG	4	>25	0	1	0
Caryophanon latum L	H. Mayer	*Cla*I	AT↓CGAT	15	2	0	0	1
Caryophanon latum	ATCC 15219	*Clm*I (*Hae*III)	GGCC	>50	>50	19	11	22
		*Clm*II (*Ava*II)	GG(^A_T)CC	>17	>30	6	1	8
Caryophanon latum	DSM 484	*Clt*I (*Hae*III)	GG↓CC	>50	>50	19	11	22
Caryophanon latum RII	H. Mayer	*Clu*I	?	>20	?	?	?	?
Caryophanon latum H7	W.C. Trentini	*Cal*I	?	14	?	?	?	?
Caulobacter crescentus CB-13	R.J. Syddall	*Ccr*I	?	1	>10	0	1	1
		*Ccr*II (*Xho*I)	CTCGAG	1	6	0	1	0
Chloroflexus aurantiacus	A. Bingham	*Cau*I (*Ava*II)	GG(^A_T)CC	>30	>30	6	1	8
		*Cau*II	CC↓(^C_G)GG	>30	>30	0	1	10
Chromatium vinosum	G.C. Grosveld	*Cvn*I (*Sau*I)	CC↓TNAGG	>10	>15	0	1	0
Chromobacterium violaceum	ATCC 12472	*Cvi*I	?	?	?	?	?	?
Citrobacter freundii	A.A. Janulaitis	*Cfr*I	Py↓GGCCPu	>25	>35	0	2	6
Clostridium formicoaceticum	ATCC 23439	*Cfo*I (*Hha*I)	GCGC	>50	>50	2	18	31
Clostridium pasteurianum	NRCC 33011	*Cpa*I (*Mbo*I)	GATC	>50[d]	>50	8	0	22
Corynebacterium humiferum	ATCC 21108	*Chu*I (*Hind*III)	AAGCTT	6	11	6	0	1
		*Chu*II (*Hind*II)	GTPyPuAC	34	>20	7	13	2
Corynebacterium petrophilum	ATCC 19080	*Cpe*I (*Bcl*I)	TGATCA	7[d]	5	1	0	0
Cystobacter velatus Plv9	H. Reichenbach	*Cve*I	?	?	?	?	?	?
Desulfovibrio desulfuricans Norway strain	H. Peck	*Dde*I	C↓TNAG	>50	>50	19	14	8
		*Dde*II (*Xho*I)	CTCGAG	1	6	0	1	0
Desulfovibrio desulfuricans	ATCC 27774	*Dds*I (*Bam*HI)	GGATCC	5	3	1	0	1
Diplococcus pneumoniae	S. Lacks	*Dpn*I	GA*↓TC	only cleaves methylated DNA				
Diplococcus pneumoniae	S. Lacks	*Dpn*II (*Mbo*I)	GATC	>50[d]	>50	8	0	22
Enterobacter aerogenes	P.R. Whitehead	*Eae*I (*Cfr*I)	Py↓GGCCPu	>25	>35	0	2	6
Enterobacter cloacae	H. Hartmann	*Ecl*I	?	14	?	?	?	?
		*Ecl*II (*Eco*RII)	CC(^A_T)GG	>35[d]	>35	16	2	6
Enterobacter cloacae	DSM 30056	*Eca*I (*Bst*EII)	G↓GTNACC	11	8	0	0	0
		*Eca*II (*Eco*RII)	CC(^A_T)GG	>35[d]	>35	16	2	6
Enterobacter cloacae	DSM 30060	*Ecc*I (*Sac*II)	CCGCGG	4	>25	0	1	0
Escherichia coli J62 pLG74	L.I. Glatman	*Eco*RV	GATAT↓C	14	8	1	0	1
Escherichia coli RY13	R.N. Yoshimori	*Eco*RI	G↓AATTC	5	5	1	0	1
		*Eco*RI'	PuPuA↓TPyPy	>10	>10	24	16	15
Escherichia coli R245	R.N. Yoshimori	*Eco*RII	↓CC(^A_T)GG	>35[d]	>35	16	2	6
Escherichia coli B	W. Arber	*Eco*B	TGA(N)$_8$TGCT	Type I	1	0	0	
Escherichia coli K	M. Meselson	*Eco*K	AAC(N)$_6$GTGC	Type I	0	0	2	
Escherichia coli (PI)	K. Murray	*Eco*PI	AGACC	Type III	4	7	4	
Escherichia coli P15	W. Arber	*Eco*P15	CAGCAG	Type III	12	5	7	

Microorganism	Source	Enzyme[a]	Sequence[b]	Number of cleavage sites[c]				
				λ	Ad2	SV40	φX	pBR
Flavobacterium okeanokoites	IFO 12536	*Fok*I	GGATG (9/13)	>50	>50	11	8	6
Fremyella diplosiphon	PCC 7601	*Fdi*I (*Ava*II)	G↓G(^A_T)CC	>17	>30	6	1	8
		*Fdi*II (*Mst*I)	TGC↓GCA	>10	>15	0	1	4
Fusobacterium nucleatum A	M. Smith	*Fnu*AI (*Hin*fI)	G↓ANTC	>50	>50	10	21	10
		*Fnu*AII (*Mbo*I)	GATC	>50[d]	>50	8	0	22
Fusobacterium nucleatum C	M. Smith	*Fnu*CI (*Mbo*I)	↓GATC	>50[d]	>50	8	0	22
Fusobacterium nucleatum D	M. Smith	*Fnu*DI (*Hae*III)	GG↓CC	>50	>50	19	11	22
		*Fnu*DII	CG↓CG	>50	>50	0	14	23
		*Fnu*DIII (*Hha*I)	GCG↓C	>50	>50	2	18	31
Fusobacterium nucleatum E	M. Smith	*Fnu*EI (*Mbo*I)	↓GATC	>50[d]	>50	8	0	22
Fusobacterium nucleatum 48	M. Smith	*Fnu*48I	?	>50	?	?	>10	?
Fusobacterium nucleatum 4H	M. Smith	*Fnu*4HI	GC↓NGC	>50	>50	25	31	42
Gluconobacter dioxyacetonicus	IAM 1814	*Gdi*I (*Stu*I)	AGG↓CCT	5	12	7	1	0
		*Gdi*II	Py↓GGCCG	>10	>20	0	2	5
Gluconobacter dioxyacetonicus	IAM 1840	*Gdo*I (*Bam*HI)	GGATCC	5	3	1	0	1
Gluconobacter oxydans sub. *melonogenes*	IAM 1836	*Gox*I (*Bam*HI)	GGATCC	5	3	1	0	1
Haemophilus aegyptius	ATCC 11116	*Hae*I	(^A_T)GG↓CC(^A_T)	?	?	11	6	7
		*Hae*II	PuGCGC↓Py	>30	>30	1	8	11
		*Hae*III	GG↓CC	>50	>50	19	11	22
Haemophilus aphrophilus	ATCC 19415	*Hap*I	?	>30	?	?	?	?
		*Hap*II (*Hpa*II)	C↓CGG	>50	>50	1	5	26
Haemophilus gallinarum	ATCC 14385	*Hga*I	GACGC (5/10)	>50	>50	0	14	11
Haemophilus haemoglobinophilus	ATCC 19416	*Hhg*I (*Hae*III)	GGCC	>50	>50	19	11	22
Haemophilus haemolyticus	ATCC 10014	*Hha*I	GCG↓C	>50	>50	2	18	31
		*Hha*II (*Hin*fI)	GANTC	>50	>50	10	21	10
Haemophilus influenzae GU	J. Chirikjian	*Hin*GUI (*Hha*I)	GCGC	>50	>50	2	18	31
		*Hin*GUII (*Fok*I)	GGATG	>50	>50	11	8	6
Haemophilus influenzae 173	J. Chirikjian	*Hin* 173 (*Hind*III)	AAGCTT	6	11	6	0	1
Haemophilus influenzae 1056	J. Stuy	*Hin* 1056I (*Fnu*DII)	CGCG	>50	>50	0	14	22
		Hin 1056II	?	>30	>30	0	5	?
Haemophilus influenzae b (1076)	J. Stuy	*Hin*bIII (*Hind*III)	AAGCTT	6	11	6	0	1
Haemophilus influenzae c (1160)	J. Stuy	*Hin*cII (*Hind*II)	GTPyPuAC	34	>20	7	13	2
Haemophilus influenzae c (1161)	J. Stuy	*Hin*cII (*Hind*II)	GTPyPuAC	34	>20	7	13	2
Haemophilus influenzae e	A. Piekarowicz	*Hin*eI (*Hin*fIII)	CGAAT[f]	Type III		0	5	1
Haemophilus influenzae R$_b$	C.A. Hutchison	*Hin*bIII (*Hind*III)	AAGCTT	6	11	6	0	1
Haemophilus influenzae R$_c$	A. Landy	*Hin*cII (*Hind*II)	GTPyPuAC	34	>20	7	13	2
Haemophilus influenzae R$_d$	S.H. Goodgal (exo⁻mutant)	*Hind*I	CẤC	specific methylase				
		*Hind*II	GTPy↓PuẤC	34	>20	7	13	2
		*Hind*III	Ấ↓AGCTT	6	11	6	0	1
		*Hind*IV	GẤC	specific methylase				
Haemophilus influenzae R$_1$	C.A. Hutchison	*Hin*fI	G↓ANTC	>50	>50	10	21	10
		*Hin*fII (*Hind*III)	AAGCTT	6	11	6	0	1
		*Hin*fIII	CGAAT[f]	Type III		0	5	1
Haemophilus influenzae H-1	M. Takanami	*Hin*HI (*Hae*II)	PuGCGCPy	>30	>30	1	8	11
Haemophilus influenzae P$_1$	S. Shen	*Hin*P$_1$I (*Hha*I)	G↓CGC	>50	>50	2	18	31
Haemophilus influenzae S$_1$	S. Shen	*Hin*S$_1$I (*Hha*I)	GCGC	>50	>50	2	18	31
Haemophilus influenzae S$_2$	S. Shen	*Hin*S$_2$I (*Hha*I)	GCGC	>50	>50	2	18	31
Haemophilus influenzae JC9	A. Piekarowicz	*Hin*JCI (*Hind*II)	GTPy↓PuAC	34	>20	7	13	2
		*Hin*JCII (*Hind*III)	AAGCTT	6	11	6	0	1
Haemophilus parahaemolyticus	C.A. Hutchison	*Hph*I	GGTGA (8/7)[b]	>50	>50	4	9	12
Haemophilus parainfluenzae	J. Setlow	*Hpa*I	GTT↓AẤC	13	6	4	3	0
		*Hpa*II	C↓CGG	>50	>50	1	5	26
Haemophilus suis	ATCC 19417	*Hsu*I (*Hind*III)	A↓AGCTT	6	11	6	0	1
Herpetosiphon giganteus HP1023	J.H. Parish	*Hgi*AI	G(^A_T)GC(^A_T)↓C	24	>20	0	3	8

Microorganism	Source	Enzyme[a]	Sequence[b]	Number of cleavage sites[c]				
				λ	Ad2	SV40	φX	pBR
Herpetosiphon giganteus Hpg 5	H. Reichenbach	*Hgi*BI (*Ava*II)	G↓G(^A_T)CC	>17	>30	6	1	8
Herpetosiphon giganteus Hpg 9	H. Reichenbach	*Hgi*CI	G↓GPyPuCC	13	>25	1	3	9
		*Hgi*CII (*Ava*II)	G↓G(^A_T)CC	>17	>30	6	1	8
		*Hgi*CIII (*Sal*I)	G↓TCGAC	2	3	0	0	1
Herpetosiphon giganteus Hpa 2	H. Reichenbach	*Hgi*DI (*Acy*I)	GPu↓CGPyC	>14	>14	0	7	6
		*Hgi*DII (*Sal*I)	G↓TCGAC	2	3	0	0	1
Herpetosiphon giganteus Hpg 24	H. Reichenbach	*Hgi*EI (*Ava*II)	G↓G(^A_T)CC	>17	>30	6	1	8
		*Hgi*EII	ACC(N)$_6$GGT	?	?	1	1	2
Herpetosiphon giganteus Hpg 14	H. Reichenbach	*Hgi*FI	?	?	15	?	?	?
Herpetosiphon giganteus Hpa 1	H. Reichenbach	*Hgi*GI (*Acy*I)	GPu↓CGPyC	>14	>14	0	7	6
Herpetosiphon giganteus Hp 1049	J.H. Parish	*Hgi*HI (*Hgi*CI)	G↓GPyPuCC	>13	>25	1	3	9
		*Hgi*HII (*Acy*I)	GPu↓CGPyC	>14	>14	0	7	6
		*Hgi*HIII (*Ava*II)	G↓G(^A_T)CC	>17	>30	6	1	8
Herpetosiphon giganteus HFS 101	H. Foster	*Hgi*JI	?	?	?	?	?	?
		*Hgi*JII	GPuGCPy↓C	7	>35	2	0	2
Herpetosiphon giganteus Hpg 32	H. Reichenbach	*Hgi*KI	?	>18	>20	?	?	?
Klebsiella pneumoniae OK8	J. Davies	*Kpn*I	GGTAC↓C	2	8	1	0	0
Mastigocladus laminosus	CCAP 1447/1	*Mla*I (*Asu*II)	TT↓CGAA	7	1	0	0	0
Microbacterium thermo-sphactum	ATCC 11509	*Mth*I (*Mbo*I)	GATC	>50[d]	>50	8	0	22
Micrococcus luteus	IFO 12992	*Mlu*I	A↓CGCGT	?	5	0	2	0
Micrococcus radiodurans	ATCC 13939	*Mra*I (*Sac*II)	CCGCGG	4	25	0	1	0
Microcoleus species	D. Comb	*Mst*I	TGC↓GCA	>10	>15	0	1	4
		*Mst*II (*Sau*I)	CC↓TNAGG	2	7	0	0	0
Moraxella bovis	ATCC 10900	*Mbo*I	↓GATC	>50[d]	>50	8	0	22
		*Mbo*II	GAAGA (8/7)	>50	>50	16	11	11
Moraxella bovis	ATCC 17947	*Mbv*I	?	?	?	?	?	?
Moraxella glueidi LG1	J. Davies	*Mgl*I	?	?	?	?	?	?
Moraxella glueidi LG2	J. Davies	*Mgl*II	?	?	?	?	?	?
Moraxella kingae	ATCC 23331	*Mki*I (*Hind*III)	AAGCTT	6	11	6	0	1
Moraxella nonliquefaciens	ATCC 19975	*Mno*I (*Hpa*II)	C↓CGG	>50	>50	1	5	26
		*Mno*II (*Mnn*III)	?	>10	>6	3	?	?
		*Mno*III (*Mbo*I)	GATC	>50[d]	>50	8	0	22
Moraxella nonliquefaciens	ATCC 17953	*Mni*I	CCTC (7/7)	>50	>50	51	34	26
Moraxella nonliquefaciens	ATCC 17954	*Mnn*I (*Hind*II)	GTPyPuAC	34	20	7	13	2
		*Mnn*II (*Hae*III)	GGCC	>50	>50	19	11	22
		*Mnn*III	?	>10	>6	3	?	?
		*Mnn*IV (*Hha*I)	GCGC	>50	>50	2	18	31
Moraxella nonliquefaciens	ATCC 19996	*Mni*I (*Hae*III)	GGCC	>50	>50	19	11	22
		*Mni*II (*Hpa*II)	CCGG	>50	>50	1	5	26
Moraxella osloensis	ATCC 19976	*Mos*I (*Mbo*I)	GATC	>50[d]	>50	8	0	22
Moraxella phenylpyruvica	ATCC 17955	*Mph*I (*Eco*RII)	CC(^A_T)GG	>35[d]	>35	16	2	6
Moraxella species[h]	R.J. Roberts	*Msp*I (*Hpa*II)	C↓CGG	>50	>50	1	5	26
Myxococcus stipitatus Mxs2H	H. Reichenbach	*Msi*I (*Xho*I)	CTCGAG	1	6	0	1	0
		*Msi*II	?	?	?	?	?	?
Myxococcus virescens V-2	H. Reichenbach	*Mvi*I	?	1	?	?	?	?
		*Mvi*II	?	?	?	?	?	?
Neisseria caviae	NRCC 31003	*Nca*I (*Hinf*I)	GANTC	>50	>50	10	21	10
Neisseria cinerea	NRCC 31006	*Nci*I (*Cau*II)	CC↓(^C_G)GG[g]	>15	>15	0	1	10

Microorganism	Source	Enzyme[a]	Sequence[b]	Number of cleavage sites[c]				
				λ	Ad2	SV40	φX	pBR
Neisseria denitrificans	NRCC 31009	*Nde*I	CATATG	?	?	2	0	1
		*Nde*II (*Mbo*I)	GATC	>50[d]	>50	8	0	22
Neisseria flavescens	NRCC 31011	*Nfl*I (*Mbo*I)	GATC	>50[d]	>50	8	0	22
		*Nfl*II	?	?	?	?	?	?
		*Nfl*III	?	?	?	?	?	?
Neisseria gonorrhoea	G. Wilson	*Ngo*I (*Hae*II)	PuGCGCPy	>30	>30	1	8	11
Neisseria gonorrhoea	CDC 66	*Ngo*II (*Hae*III)	GGCC	>50	>50	19	11	22
Neisseria gonorrhoea KH 7764-45	L. Mayer	*Ngi*III (*Sac*II)	CCGCGG	4	>25	0	1	0
Neisseria mucosa	NRCC 31013	*Nmu*I (*Nae*I)	GCCGGC	2	>13	1	0	4
Neisseria ovis	NRCC 31020	*Nov*I	?	?	?	?	?	?
		*Nov*II (*Hinf*I)	GANTC	>50	>50	10	21	10
Nocardia aerocolonigenes	ATCC 23870	*Nae*I	GCC↓GGC	2	>13	1	0	4
Nocardia argentinensis	ATCC 31306	*Nar*I	GG↓CGCC	2	10	0	2	4
Nocardia blackwellii	ATCC 6846	*Nbl*I (*Pvu*I)	CGAT↓CG	3	7	0	0	1
Nocardia corallina	ATCC 19070	*Nco*I	CCATGG	6	14	3	0	0
Nocardia opaca	ATCC 21507	*Nop*I (*Sal*I)	G↓CGAC	2	3	0	0	1
		*Nop*II (*Sal*II)	?	?	?	?	?	?
Nocardia otitidis-caviarum	ATCC 14630	*Not*I	?	0	6	0	0	0
Nocardia rubra	ATCC 15906	*Nru*I	TCG↓CGA	7	6	0	2	1
Nocardia uniformis	ATCC 21806	*Nun*I	?	?	?	?	?	?
		*Nun*II (*Nar*I)	GG↓CGCC	2	14	0	2	4
Nostoc species	PCC 6705	*Nsp*BI	?	?	?	?	?	?
		*Nsp*BII	GC($_G^C$)G↓C	?	?	1	17	21
Nostoc species	PCC 7524	*Nsp*CI	PuCATG↓Py	>15	>15	2	0	4
Nostoc species	PCC 7413	*Nsp*HI (*Nsp*CI)	PuCATG↓Py	>15	>15	2	0	4
Oerskovia xanthineolytica	R. Shekman	*Oxa*I (*Alu*I)	AGCT	>50	>50	35	24	16
		*Oxa*II	?	?	?	?	?	?
Proteus vulgaris	ATCC 13315	*Pvu*I	CGAT↓CG	3	7	0	0	1
		*Pvu*II	CAG↓CTG	15	22	3	0	1
Providencia alcalifaciens	ATCC 9886	*Pal*I (*Hae*III)	GGCC	>50	>50	19	11	22
Providencia stuartii 164	J. Davies	*Pst*I	CTGCA↓G	18	25	2	1	1
Pseudoanabaena species	ATCC 27263	*Psp*I (*Asu*I)	GGNCC	>30	>30	11	2	15
Pseudomonas aeruginosa	G.A. Jacoby	*Pae*R7	?	1	5	0	0	0
Pseudomonas facilis	M. Van Montagu	*Pfa*I (*Mbo*I)	GATC	>50[d]	>50	8	0	22
Pseudomonas maltophila	D. Comb	*Pma*I (*Pst*I)	CTGCAG	18	25	2	1	1
Rhizobium leguminosarum 300	J. Beringer	*Rle*I	?	6	>10	?	?	?
Rhizobium lupini #1	W. Heumann	*Rlu*I	?	1	8	?	?	?
Rhizobium meliloti	J.L. Denarie	*Rme*I	?	8	>10	?	?	?
Rhodospirillum rubrum	J. Chirikjian	*Rrb*I	?	?	4	5	1	?
Rhodopseudomonas sphaeroides	R. Lascelles	*Rsp*I (*Pvu*I)	CGATCG	3	7	0	0	1
Rhodopseudomonas sphaeroides	S. Kaplan	*Rsh*I (*Pvu*I)	CGAT↓CG	3	7	0	0	1
Rhodopseudomonas sphaeroides	S. Kaplan	*Rsa*I	GT↓AC	>50	>50	11	11	3
Rhodopseudomonas sphaeroides	S. Kaplan	*Rsr*I (*Eco*RI)	GAATTC	5	5	1	0	1
Salmonella infantis	A. deWaard	*Sin*I (*Ava*II)	GG($_T^A$)CC	>17	>30	6	1	8
Serratia marcescens S$_b$	C. Mulder	*Sma*I	CCC↓GGG	3	12	0	0	0
Serratia species SAI	B. Torheim	*Ssp*I	?	?	?	?	?	?
Sphaerotilus natans C	A. Pope	*Sna*I	GTATAC	2	?	0	0	1
Spiroplasma citri ASP2	M.A. Stephens	*Sci*NI (*Hha*I)	G↓CGC	>50	>50	2	18	31
Staphylococcus aureus 3A	E.E. Stobberingh	*Sau*3A (*Mbo*I)	↓GATC	>50[d]	>50	8	0	22
Staphylococcus aureus PS96	E.E. Stobberingh	*Sau*961 (*Asu*I)	G↓GNCC	>30	>30	11	2	15
Staphylococcus saprophyticus	ATCC 13518	*Ssa*I	?	>10	?	?	?	?
Streptococcus cremoris F	C. Daly	*Scr*FI	CCNGG	>50	>50	16	3	16
Streptococcus faecalis var. *zymogenes*	R. Wu	*Sfa*I (*Hae*III)	GG↓CC	>50	>50	19	11	22
Streptococcus faecalis GU	J. Chirikjian	*Sfa*GUI (*Hpa*II)	CCGG	>50	>50	1	5	26
Streptococcus faecalis ND 547	D. Clewell	*Sfa*NI	GCATC (5/9)	>50	>30	6	12	22

Microorganism	Source	Enzyme[a]	Sequence[b]	Number of cleavage sites[c]				
				λ	Ad2	SV40	φX	pBR
Streptomyces achromogenes	ATCC 12767	*Sac*I	GAGCT↓C	2	7	0	0	0
		*Sac*II	CCGC↓GG	4	>25	0	1	0
		*Sac*III	?	>100	>100	?	?	?
Streptomyces albus	CM1 52766	*Sal*PI (*Pst*I)	CTGCA↓G	18	25	2	1	1
Streptomyces albus subspecies *pathocidicus*	KCC S0166	*Spa*I (*Xho*I)	CTCGAG	1	6	0	1	0
Streptomyces albus G	J.M. Ghuysen	*Sal*I	G↓TCGAC	2	3	0	0	1
		*Sal*II	?	>20	?	?	?	?
Streptomyces aureofaciens IKA 18/4	J. Timko	*Sau*I	CC↓TNAGG	2	7	0	0	0
Streptomyces bobili	ATCC 3310	*Sbo*I (*Sac*II)	CCGCGG	4	>25	0	1	0
Streptomyces caespitosus	H. Takahashi	*Sca*I	AGTACT	6	5	0	0	1
Streptomyces cupidosporus	KCC S0316	*Scu*I (*Xho*I)	CTCGAG	1	6	0	1	0
Streptomyces exfoliatus	KCC S0030	*Sex*I (*Xho*I)	CTCGAG	1	6	0	1	0
		*Sex*II	?	2	?	?	?	?
Streptomyces fradiae	ATCC 3355	*Sfr*I (*Sac*II)	CCGCGG	4	>25	0	1	0
Streptomyces ganmycicus	KCC S0759	*Sga*I (*Xho*I)	CTCGAG	1	6	0	1	0
Streptomyces goshikiensis	KCC S0294	*Sgo*I (*Xho*I)	CTCGAG	1	6	0	1	0
Streptomyces griseus	ATCC 23345	*Sgr*I	?	0	7	0	?	?
Streptomyces hygroscopicus	T. Yamaguchi	*Shy*TI	?	2	?	?	?	?
Streptomyces hygroscopicus	F. Walter	*Shy*I (*Sac*II)	CCGCGG	4	>25	0	1	0
Streptomyces lavendulae	ATCC 8664	*Sla*I (*Xho*I)	C↓TCGAG	1	6	0	1	0
Streptomyces luteoreticuli	KCC S0788	*Slu*I (*Xho*I)	CTCGAG	1	6	0	1	0
Streptomyces oderifer	ATCC 6246	*Sod*I	?	?	?	?	?	?
		*Sod*II	?	?	?	?	?	?
Streptomyces phaeo-chromogenes	F. Bolivar	*Sph*I	GCATG↓C	4	9	2	0	1
Streptomyces stanford	S. Goff, A. Rambach	*Sst*I (*Sac*I)	GAGCT↓C	2	7	0	0	0
		*Sst*II (*Sac*II)	CCGC↓GG	4	>25	0	1	0
		*Sst*III (*Sac*III)	?	>100	>100	?	?	?
		*Sst*IV (*Bcl*I)	TGATCA	7	5	1	0	0
Streptomyces tubercidicus	H. Takahashi	*Stu*I	AGG↓CCT	5	12	7	1	0
Streptoverticillium flavopersicum	Upjohn UC 5066	*Sfl*I (*Pst*I)	CTGCA↓G	18	25	2	1	1
Thermoplasma acidophilum	D. Searcy	*Tha*I (*Fnu*DII)	CG↓CG	>50	>50	0	14	23
Thermopolyspora glauca	ATCC 15345	*Tgl*I (*Sac*II)	CCGCGG	4	>25	0	1	0
Thermus aquaticus YTI	J.I. Harris	*Taq*I	T↓CGA*Ǎ	>50	>50	1	10	7
		*Taq*II	?	>30	>30	4	6	?
Thermus aquaticus	S.A. Grachev	*Taq*XI (*Eco*RII)	C*Č↓(^A_T)GG	>35[d]	>35	16	2	6
Thermus flavus AT62	T. Oshima	*Tfl*I (*Taq*I)	TCGA	>50	>50	1	10	7
Thermus thermophilus HB8	T. Oshima	*Tth*HB8 I (*Taq*I)	TCG*Ǎ	>50	>50	1	10	7
Thermus thermophilus strain 23	T. Oshima	*Ttr*I (*Tth* 111 I)	GACNNNGTC	2	>12	0	0	1
Thermus thermophilus strain 110	T. Oshima	*Tte*I (*Tth* 111 I)	GACNNNGTC	2	>12	0	0	1
Thermus thermophilus strain 111	T. Oshima	*Tth* 111 I	GACN↓NNGTC	2	>12	0	0	1
		Tth 111 II	CAAPuCA	>30	>30	12	11	5
		Tth 111 III	?	?	?	?	?	?
Tolypothrix tenuis	W. Seigelman	*Ttn*I (*Hae*III)	GGCC	>50	>50	19	11	22
Xanthomonas amaranthicola	ATCC 11645	*Xam*I (*Sal*I)	GTCGAC	2	3	0	0	1
Xanthomonas badrii	ATCC 11672	*Xba*I	T↓CTAGA	1[d]	4	0	0	0
Xanthomonas holcicola	ATCC 13461	*Xho*I	C↓TCGAG	1	6	0	1	0
		*Xho*II	Pu↓GATCPy	>20	>20	3	0	8
Xanthomonas malvacearum	ATCC 9924	*Xma*I (*Sma*I)	C↓CCGGG	3	12	0	0	0
		*Xma*II (*Pst*I)	CTGCAG	18	25	2	1	1
		*Xma*III	C↓GGCCG	2	10	0	0	1
Xanthomonas manihotis 7AS1	B-C. Lin	*Xmn*I	GAANNNNTTC	>11	?	0	3	2
Xanthomonas nigromaculans	ATCC 23390	*Xni*I (*Pvu*I)	CGATCG	3	7	0	0	1
Xanthomonas oryzae	M. Ehrlich	*Xor*I (*Pst*I)	CTGCAG	18	25	2	1	1
		*Xor*II (*Pvu*I)	CGAT↓CG	3	7	0	0	1
Xanthomonas papavericola	ATCC 14180	*Xpa*I (*Xho*I)	C↓TCGAG	1	6	0	1	0

NOTES

[a]When two enzymes recognize the same sequence, i.e., are isoschizomers, the prototype (i.e., the first example isolated) is indicated in parentheses.

[b]Recognition sequences are written from 5′ → 3′, only one strand being given, and the point of cleavage, when known, is indicated (ˆ). For example, GˆGATCC is an abbreviation for:

$$5' \text{ G}\hat{~}\text{G A T C C } 3'$$
$$3' \text{ C C T A G}\hat{~}\text{G } 5'$$

For enzymes such as *Hga*I, *Mbo*II etc, which cleave away from their recognition sequence the sites of cleavage are indicated in parentheses. For example *Hga*I GACGC(5/10) indicates cleavage as shown below

$$5'\text{GACGCNNNNN}\hat{~}\qquad 3'$$
$$3'\text{CTGCGNNNNNNNNNN}\hat{~}5'$$

In all cases the recognition sequences are oriented so that the cleavage sites lie on their 3′ side.

Note that cleavage by *Hpb*I can vary. A G/C base pair at nucleotide 8 favors cleavage after 9, not 8, bases.

Bases appearing in parentheses signify that either base may occupy that position in the recognition sequence. Thus, *Acc*I cleaves the sequences GTAGAC, GTATAC, GTACGAC, and GTCTAC. Where known, the base modified by the corresponding specific methylase is indicated by an asterisk. $\overset{*}{\text{A}}$ is N^6-methyladenosine. $\overset{*}{\text{C}}$ is 5-methylcytosine.

[c]These columns indicate the frequency of cleavage by the various specific endonucleases on bacteriophage lambda DNA ( ), adenovirus-2 DNA (Ad2), simian virus 40 DNA (SV40), phiX174 Rf DNA (ØX) and pBR322 DNA (pBR). In the latter three cases, the sites were checked by computer search of the published sequences.

[d]In most *E. coli* strains, bacteriophage lambda DNA is partially modified against the action of *Atu*BI, *Atu*II, *Atu*CI, *Bcl*I, *Bss*GII, *Bst*GI, *Bst*GII, *Bst*XI, *Bst*EIII, *Cpa*I, *Cpe*I, *Dpn*II, *Eca*II, *Ecl*II, *Eco*RII, *Fnu*AII, *Fnu*CI, *Mbo*I, *Mno*III, *Mos*I, *Mph*I and *Nde*II. *Xba*I also rarely gives complete digestion presumably because the *Xba*I recognition sequence overlaps a *dam* recognition site. It should be noted that *Mth*I, *Fnu*EI, *Pfa*I and *Sau*3A are not inhibited by *dam* methylation; *Bst*NI, *Tao*XI and *Apy*I are not inhibited by *mec* methylation.

[e]*Eco*PI, *Eco*P15, *Hine*I and *Hinf*III have characteristics intermediate between those of the Type I and Type II restriction endonucleases. They are designated Type III in accordance with the suggestion of Kauc and Piekarowicz (98).

[f]Both *Hinf*III and *Hine*I cleave about 25 bases 3′ of the recognition sequence.

[g]*Nci*I leaves termini carrying a 3′-phosphate group (87).

[h]*Acinetobacter calcoaceticus* has recently been identified as a *Flavobacterium* species and *Moraxella* species has recently been identified as *Acinetobacter calcoaceticus* (J. Kane, unpublished observations). The old names will remain associated with these strains to avoid confusion.

References

Achtman, M., H. Willets, and A. J. Clark (1971). Beginning a genetic analysis of conjugational transfer determined by the F-factor in *Escherichia coli* by isolating and characterization of transfer-deficient mutants. *J. Bacteriol.* **106**:529.

Bachmann, B. J. (1972). Pedigrees of some mutant strains of *Escherichia coli* K-12. *Bacter. Rev.* **36**:525.

Bachmann, B. J. and K. B. Low (1980). Linkage map of *Escherichia coli* K-12. (6th ed.) *Microbiol. Rev.* **44**:1.

Bailey, J. M. and N. Davidson (1976). Methylmercury as a reversible denaturing agent for agarose gel electrophoresis. *Anal. Biochem.* **70**:75.

Bernard, H. U. and D. R. Helinski (1980). Bacterial plasmid cloning vehicles. In *Genetic Engineering* (J. K. Setlow and A. E. Hollaender, eds.). Plenum Press, New York, **3**:133.

Blattner, F. R., B. G. Williams, A. E. Blechl, K. Denniston-Thompson, H. E. Faber, L. A. Furlong, D. J. Grunwald, D. O. Kiefer, D. D. Moore, E. L. Sheldon, and O. Smithies (1977). Charon phages: Safer derivatives of bacteriophage lambda for DNA cloning. *Science* **196**:161.

Bolivar, F. (1978). Construction and characterization of new cloning vehicles. III. Derivatives of plasmid pBR322 carrying unique *Eco*RI sites for selection of *Eco*RI generated recombinant DNA molecules. *Gene* **4**:121.

Calos, M. (1978). DNA sequence for a low-level promoter of the *lac* repressor gene and an "up" promoter mutation. *Nature* **274**:762–765.

Collins, J., M. Johnson, P. Jorgensen, P. Valentine-Hansen, H. O. Karlstrom, F. Gautier, W. Lendenmaier, H. Moyer, and B. M. Sjoberg (1978). Expression of plasmid genes: ColEl and derivatives. In *Microbiology* (D. Schlessinger, ed.). American Society of Microbiology, Washington, D.C., p. 150.

Crosa, J. H. and S. Falkow (1981). Plasmids. In *Manual of Methods for General Bacteriology* (P. Gerhardt et al., eds.). American Society for Microbiology, Washington, D.C., p. 266.

Davis, R. W., D. Botstein, and J. R. Roth (1980). *Advanced Bacterial Genetics*. Cold Spring Harbor Laboratory, Cold Spring Harbor, New York.

Demerec, M., E. A. Adelberg, A. J. Clark, and P. E. Hartman (1966). A proposal for uniform nomenclature in bacterial genetics. *Genetics* **54**:61.

Dugaiczyk, A., H. W. Boyer, and H. M. Goodman (1975). Ligation of *Eco*RI endonuclease-generated DNA fragments into linear and circular structures. *J. Mol. Biol.* **96**:171.

Ericksson, S., B. -M. Sjoberg, S. Hahne, and O. Karlstrom (1977). Ribonucleotide diphosphate reductase from *Escherichia coli*. An immunological assay and a novel purification from an overproducing strain lysogenic for phage λd*nrd*. *Biol. Chem.* **252**:6132.

Gillam, S. and M. Smith (1979). Site-specific mutagenesis using oligonucleotide primers. *Gene* **8**: 81; 99.

Hanahan, D. (1983). Studies on transformation of *Escherichia coli* with plasmids. *J. Mol. Biol.* **166**: 557.

Heynecker, H. L., J. Shine, H. M. Goodman, H. W. Boyer, I. Rosenberg, R. E. Dickerson, S. A. Narange, K. Itakura, S. Ling, and A. D. Riggs (1976). Synthetic *lac* operator DNA is functional *in vivo*. *Nature* **263**:748.

Howarth, A. J., R. C. Gardner, J. Messing, and R. J. Shepherd (1981). Nucleotide sequences of naturally occurring deletion mutants of Cauliflower mosaic virus. *Virology* **112**:768.

Hutchison, C. A. III, S. Phillips, M. H. Edgell, S. Gillam, P. Johnke, and M. Smith (1978). Mutagenesis at a specific position in a DNA sequence. *J. Biol. Chem.* **253**:6551.

Ish-Horowitz, D. and J. F. Burke (1981). Rapid and efficient cosmid cloning. *Nucl. Acids Res.* **9**:2989.

Koch, A. (1981). In *Manual of Methods for General Bacteriology* (P. Gerhardt *et al.*, eds.). American Society for Microbiology, Washington, D.C., **179**:207.

Kornberg, A. (1980). *DNA Replication*. W. H. Freeman & Co., San Francisco.

Landy, A., E. Olchowski, W. Ross, and G. Reiness (1974). Isolation of a functional *lac* regulatory region. *Mol. Gen. Genet.* **133**:273.

Langley, K. E., M. E. Villarejo, A. V. Fowler, P. J. Zamenhof, and T. Zabin (1975). Molecular basis of β-galactosidase α complementation. *Proc. Natl. Acad. Sci.* **72**:1254.

Lapage, S. P., J. E. Skelton, T. G. Mitchell, and A. R. Mackenzie (1970). Culture collections and the preservation of bacteria. In *Methods in Microbiology* (J. R. Norris and D. W. Ribbons, eds.). Academic Press, London, **3A**:135.

Leder, P., D. Tiemeier, and L. Enquist (1977). EK2 derivatives of bacteriophage lambda useful in the cloning of DNA from higher organisms: The λgtWES system. *Science* **196**:175.

Maniatis, T., E. F. Fritsch, and J. Sambrook (1982). *Molecular Cloning*. Cold Spring Harbor Laboratory, Cold Spring Harbor, New York.

Marmur, J. (1963). A procedure for the isolation of deoxyribonucleic acid from microorganisms. *Methods in Enzymology* **6**:726.

McMaster, G. K. and C. G. Carmichael (1977). Analysis of single- and double-stranded nucleic acids on polyacrylamide and agarose gels by using glyoxal and acridine orange. *Proc. Natl. Acad. Sci. U.S.A.* **74**:4335.

Messing, J. (1979). A multipurpose cloning system based on the single-stranded DNA bacteriophage M13. *Recombinant DNA Technical Bulletin* **2**:43.

Messing, J. (1983). New M13 vectors for cloning. In *Methods in Enzymology* (R. Wu, ed.). Academic Press, New York, **101**:20.

Messing, J., B. Gronenborn, B. Müller-Hill, and P. H. Hofschneider (1977). Filamentous coliphage M13 as a cloning vehicle: Insertion of a *Hin*dII fragment of the *lac* regulatory region in M13 replicative form *in vitro*. *Proc. Natl. Acad. Sci. U.S.A.* **74**:3642.

Messing, J. and J. Vieira (1982). A new pair of M13 vectors for selecting either strand of double-digest restriction fragments. *Gene* **19**:269.

Miller, J. H. (1972). *Experiments in Molecular Genetics.* Cold Spring Harbor Laboratory, Cold Spring Harbor, New York.

Müller-Hill, B., L. Crapo, and W. Gilbert (1968). Mutants that make more *lac* repressor. *Proc. Natl. Acad. Sci. U.S.A.* **59**:1259.

Old, R. W. and S. B. Primrose (1982). *Principles of Gene Manipulation.* (2nd ed.). University of California Press, Berkeley.

Parish, J. (1972). *Principles and Practice of Experiments with Nucleic Acids.* Longman Group, Ltd., London.

Ullmann, A., F. Jacob, and J. Monod (1967). Characterization by *in vitro* complementation of a peptide corresponding to an operator-proximal segment of the β-galactosidase structural gene of *Escherichia coli. J. Mol. Biol.* **24**:339.

Vieira, J. and J. Messing (1982). The pUC plasmids, a M13mp7 derived system for insertion mutagenesis and sequencing with synthetic universal primers. *Gene* **19**:259.

Wu, R. (ed.) (1979). Recombinant DNA. In *Methods in Enzymology.* Academic Press, New York, **68**:155.

Index

Note: Page references in italics denote figures or tables.